TESTIMONIALS

"*The Crisis Is Here* beautifully balances an intersectional, systems lens on the climate crisis with tactical information on what to do when it hits home for you and your loved ones. The book covers a wide range of those inevitable effects of climate change in your daily life, from wildfires and heat waves to floods and water contamination. I'm thrilled that Sam balances smart climate information sharing with a lot of sensible, and approachable, preparatory advice."

– Candice Ammori, Founder, Climate Vine

"*The Crisis Is Here* gave me exactly what I needed—clear, practical guidance on navigating the health impacts of climate change without panic or overwhelm. It breaks down complex topics in a straightforward way and has equipped me with tangible steps to adapt in an unpredictable world. As someone who thrives on checklists and planners, I now have a family evacuation plan and a well-organized system for storing essential items. It's not just informative, it's genuinely empowering. Gaining a deeper understanding of the challenges and having concrete strategies for myself and my community has brought me real peace of mind. At HAYVN, sustainability and social impact are core values, foundational pillars that shape everything we do. So discovering a resource that was actually written by someone within our community and aligns so closely with our mission has been incredibly meaningful."

– Felicia Rubinstein, Founder, Chief Collaborator, HAYVN Coworking

"With his comprehensive yet digestible volume, Sam Cherubin has made great strides in bringing to light the numerous and meaningful threats and health hazards with which unchecked climate change will force us all to contend to one degree or another. *The Crisis Is Here: Protecting Your Health in a Changing World* is a great resource for anyone who feels the need to become more informed about this critical issue and wishes to learn more about actions we can take as individuals and communities in order to help ensure the health of our planet and future generations."

– Larry Leibowitz, MD, Family Physician

"I first heard Sam Cherubin speak at a Foresight workshop, and what struck me then is exactly what stands out in *The Crisis Is Here*: clarity, calm urgency, and an ability to connect the dots in a way that stays with you. Climate change and public health are often treated as separate challenges, but Sam weaves them together with both scientific rigor and heart. He doesn't just explain what's happening—he helps you grasp why it matters and what you can do about it. This book gave me language, insight, and perspective I've already used in conversations with skeptics and armchair critics alike. Whether you work in public health, advocate for climate solutions, or are simply trying to make sense of an increasingly volatile world, *The Crisis Is Here* will leave you better informed and ready to act. This isn't a book about despair; it's a book about agency."

– Praniti Maini, MBA, Corporate Strategy and Foresight in Healthcare

The Crisis
Is Here

Protecting Your Health
in a Changing World

SAM CHERUBIN

For permission requests, write to the publisher, addressed "Attention: Permissions Coordinator," at the address below.

Publish Your Purpose
141 Weston Street, #155
Hartford, CT, 06141

PYP **Publish**
Your Purpose

The opinions expressed by the Author are not necessarily those held by Publish Your Purpose.

"Climate Cat"™ and the Climate Cat™ logo are trademarks of Sam Cherubin.

Ordering Information: Quantity sales and special discounts are available on quantity purchases by corporations, associations, and others. For details, contact the author at sam@samcherubin.com.

Edited by: Lily Capstick
Cover design by: Nelly Murariu
Typeset by: Nelly Murariu
Illustrations by: Daniel Dundin
Author Photo by: Tom Butler

ISBN: 979-8-88797-181-0 (hardcover)
ISBN: 979-8-88797-182-7 (paperback)
ISBN: 979-8-88797-183-4 (ebook)

Library of Congress Control Number: 2025910398

First edition, January 2026

The information contained within this book is strictly for informational purposes. The material may include information, products, or services by third parties. As such, the Author and Publisher do not assume responsibility or liability for any third-party material or opinions. The publisher is not responsible for websites (or their content) that are not owned by the publisher. Readers are advised to do their own due diligence when it comes to making decisions.

Publish Your Purpose is a hybrid publisher of non-fiction books. Our mission is to elevate the voices often excluded from traditional publishing. We intentionally seek out authors and storytellers with diverse backgrounds, life experiences, and unique perspectives to publish books that will make an impact in the world. Do you have a book idea you would like us to consider publishing? Please visit PublishYourPurpose.com for more information.

DISCLAIMER

1. The information in this book is true and complete to the best of the author's knowledge. All advice and recommendations are given without guarantee by the author or publisher. The author and publisher disclaim all liability for any use of this information.

2. The information in this book is provided for general informational and educational purposes only. The author makes no representations or warranties, express or implied, about the completeness, accuracy, reliability, suitability, or availability of any information, products, services, or related graphics contained in this book. Any use of this information is at the reader's own risk.

3. The information in this book is not intended as medical advice and should not be treated as such. Do not substitute this information for the medical advice of physicians. The information is general in nature and provided to inform readers about healthcare topics. Always consult a qualified physician regarding your individual needs or any symptoms that may require medical diagnosis or attention.

4. The advice and strategies in this book may not be suitable for every situation. This work is sold with the understanding that neither the author nor the publisher is responsible for any outcomes resulting from the use of its contents.

5. This work may reference information, products, or services provided by third parties. The author and publisher do not assume responsibility or liability for any third-party materials, content, or opinions. The publisher is not responsible for websites (or their content) that are not owned or controlled by the publisher.

6. Readers are advised to perform their own due diligence before relying on any information presented in this book when making personal, professional, financial, or health-related decisions.

DISCLAIMER

1. The information in this book is true and complete to the best of the author's knowledge. All advice and recommendations are given without guarantee by the author or publisher. The author and publisher disclaim all liability for any use of this information.

2. The information in this book is provided for general informational and educational purposes only. The author makes no representations or warranties, express or implied, about the completeness, accuracy, reliability, suitability, or availability of any information, products, services, or related graphics contained in this book. Any use of this information is at the reader's own risk.

3. The information in this book is not intended as medical advice and should not be treated as such. Do not substitute this information for the medical advice of physicians. The information is general in nature and provided to inform readers about healthcare topics. Always consult a qualified physician regarding your individual needs or any symptoms that may require medical diagnosis or attention.

4. The advice and strategies in this book may not be suitable for every situation. This work is sold with the understanding that neither the author nor the publisher is responsible for any outcomes resulting from the use of its contents.

5. This work may reference information, products, or services provided by third parties. The author and publisher do not assume responsibility or liability for any third-party materials, content, or opinions. The publisher is not responsible for websites (or their content) that are not owned or controlled by the publisher.

6. Readers are advised to perform their own due diligence before relying on any information presented in this book when making personal, professional, financial, or health-related decisions.

DEDICATION

First and always, to my wife Laura.

Thank you for your constant love, patience, and belief in me.

This journey would not have been possible without you.

CONTENTS

GLOSSARY

ADDITIONAL RESOURCES

The Interconnected Crisis: Protecting Nature to Safeguard Our Health and Future

By Samantha Taylor

The climate crisis is no longer a future threat. It is here.

From choking wildfires in California and Canada to record heatwaves across Europe and devastating floods in Libya and Vermont, climate change is no longer abstract. It threatens our health, undermines access to food and clean water, and drives millions into poverty.

As climate shocks intensify, they strain healthcare systems, destabilize economies, and deepen existing inequalities, especially between the Global North and South. One essential truth is becoming increasingly clear: the health of people is inseparable from the health of our planet. And more specifically, from the health of our ecosystems: our forests, wetlands, rivers, coral reefs, and the air we breathe.

Climate change is already costing us. In the United States alone, the public health costs of climate-related events, from respiratory diseases to heat-related illnesses and infectious disease outbreaks, are estimated at over $800 billion annually. These are not distant projections. They are real, measurable impacts on our hospitals, our homes, and the vulnerable populations who bear the brunt of a warming world.

This growing fragility isn't just environmental; it's deeply personal. As the founder of Reputation Dynamics, I've worked for years across sectors—philanthropy, corporate sustainability, global NGOs—to advance initiatives that protect nature while improving human outcomes. And I've learned that no single institution, government, or corporation can address this crisis alone. We must act collectively, as families, communities, and global citizens, to restore the balance between people and the planet.

Just as ecosystems are interconnected, so too are we.

Our family units, support networks, schools, and neighborhoods form micro-ecosystems that rely on care, communication, and trust. When disaster strikes, whether a storm, a power outage, or a health crisis, it is often neighbors, not institutions, who come to the rescue first. Rebuilding climate resilience means strengthening these relational networks while protecting the ecosystems that sustain us.

This is why this book is so timely, and so vital.

The Crisis Is Here is not just a warning, but a guide. It offers practical, empowering tools for families to prepare for, respond to, and recover from the growing health threats posed by climate change and natural disasters. It invites us to reconnect—to the natural world, and to one another.

The solutions don't only live in policy or technology. They live in the way we raise our children to care for the Earth. In how we plant trees in our neighborhoods. In how we listen to Indigenous communities who have safeguarded ecosystems for generations. In how we choose to support one another through what's coming next.

Nature is resilient. And so are we.

There is still time to restore what we have lost, to protect what we still have, and to build a healthier, more connected world. Only by remembering that we are not separate from nature—we are nature—can we ensure a future in which people and the planet thrive together. Our survival depends on protecting the ecosystems, forests, and communities that sustain life itself.

Ultimately, this is about more than crisis response. It is about legacy. About leaving our children a world where clean air, safe water, and thriving ecosystems are not luxuries, but birthrights. The decisions we make today will echo across generations.

We must also choose optimism, in the extraordinary power of nature to protect, restore, and regenerate itself when given the chance. Protecting nature means protecting ourselves, and we can still shape a future grounded in resilience, renewal, and a deep, reciprocal connection to the ecosystems that sustain us.

Sam Taylor is the founder of Reputation Dynamics, a social and environmental advisory firm advancing climate resilience, nature-based solutions, and philanthropic innovation. She is also an artist and founder of Elephant Art Shop, a creative initiative that celebrates the intersection of art, storytelling, and wildlife conservation. Sam brings decades of experience building cross-sector partnerships to support a healthier, more equitable world for people and the planet.

PREFACE

I've always had a deep love for our planet and its ecosystems. As a child, I remember wading through the shallow waters of Rockaway Beach and watching seaweed sway like delicate green ribbons beneath the surface, clinging to rocks, and drifting with the tide. On Cape Cod, I loved the prehistoric horseshoe crabs on the shoreline, their hard shells glistening in the sunlight.

I stood still in the Sierra Mountains as a dragonfly hovered motionless, its iridescent wings catching the light like stained glass. Sandhill cranes on a golf course in West Palm Beach, their long beaks digging for insects and their calls echoing across Cypress Lakes. Even in winter, when everything seems dormant, life persists. Seagulls were perched on the frozen Connecticut River, their white feathers stark against the dark ice, waiting patiently for the water to melt.

These moments aren't just poetic; they're an entryway into a world that's alive, interconnected, and constantly changing. I'm not separate from my environment. I'm part of it, woven into the fabric of the natural world just as much as the beings I love. The oceans, forests, rivers, mountains, and sky are not distant places; this is home, full of stories and rhythms that hold me.

Every space and every species are messages and medicines. Recognizing that connection gives me a deep sense of wonder and an urgent responsibility to protect what I love.

In 2015, my friend Elsa was very insistent about climate change: how catastrophic it would be, all the ways it would impact systems and society, and whether civilization would survive. With her background in food security and organizing rural women in Brazil and Africa, she had experienced the impact of drought on people and food systems.

I'd been experiencing *weird weather* in Connecticut: ice storms, tornadoes, hurricanes, week-long power outages, storm surges, intense heat, and flooding. But I was a climate ignoramus. I didn't know *why* all this was happening. Elsa's insistence on the planet's future led me on a learning journey.

Realizations came fast.

The Earth is a gigantic interplay of living systems: forests, oceans, atmosphere, ice, heat, water, and carbon distribution. As one system degrades, the others do as well. I left it there—no advocacy, no writing, no meeting others. What can one person do? Then, the world was rocked by the onset of COVID-19.

In 2020, I was working for one of the largest health insurance companies in the world. I began putting together ideas for a satellite mapping system. This tool was designed to help predict the spread of the virus. As the pandemic unfolded, my work expanded beyond those initial solo efforts.

I wanted to understand the health impacts of climate change.

I began asking questions: What are the most significant risks to public health? What can we do to help people? Who is writing about it? What is the impact on mental health?

I continued digging deeper and connecting with experts inside and outside my company. These included emergency room physicians, climate tech founders, a team from The Weather Channel, and people around the world. I interviewed doctors and nurses to understand the challenges that healthcare professionals face. I immersed myself in research and collected materials covering topics ranging from the health impacts of climate change to its effects on national security.

MY GOAL IS
TO HIGHLIGHT
THE CRISIS
AND PROVIDE
A ROADMAP
FOR PERSONAL
ACTION.

The Main Street Creamery

There was a moment when the disparity hit me. I was at an ice cream shop in Bridgeport, Connecticut, just a short drive from the Yale Program on Climate Change Communication. As I sat there licking my cone, I was struck by the disconnection between the climate information shared in academic circles, reports, and podcasts, and the everyday interactions of the people around me.

Not far away, experts were unraveling complex data about rising sea levels, extreme heat, and shifting weather patterns. But the people eating ice cream had no idea what was coming. No one was talking to them or making this global crisis feel personal or urgent in their everyday lives. They were not part of the conversation.

I wrote *The Crisis Is Here* because the book I wanted to read did not exist. Some books focus on policy, scientific projections, or broader environmental impacts, yet few provide practical advice for safeguarding our health. My goal is to highlight the crisis *and* provide a roadmap for personal action.

I needed an answer to my fundamental questions:

- ⚡ What's going to happen?
- ⚡ How will I be impacted?
- ⚡ What can I do to stay safe and healthy?
- ⚡ How can I protect myself and my family?
- ⚡ What simple steps can I take now to prepare?

When I began writing this book, I made some basic assumptions. I believed that scientific data would always be available, that government health programs would remain stable, and that weather forecasts would continue to be something we could rely on. The crisis is here. The systems we took for granted, the pillars of our society, have been shaken or have already fallen.

Government programs are being dismantled. Scientific research is under attack. We are grappling with extreme weather events that strain both our mental and physical health. I want this book to inform, uplift, and equip us for the hard reality ahead. Even now, we can still care for one another. Maintaining compassion and community as the ground shifts beneath us may be the most important thing we can do.

I grew up surrounded by the wonders of the world: monarch butterflies resting on milkweed pods, red-tailed hawks circling above highways, and mockingbirds trilling hundreds of songs. These experiences are profound because of the fragility of our ecosystems. I want to protect them and, by extension, protect us. I hope you will join me in embracing our collective responsibility to the planet and to one another. Together, we can move from unconsciousness to action. We can move from thinking, "I'm just one person," to saying, "I can and will do something." Let's begin.

INTRODUCING CLIMATE CAT

Climate Cat is a playful guide who appears throughout the book, bringing some lightheartedness to the bleak conversation about climate change and its impact on our health. He started as a joker but evolved into a teacher, mystic, and poet. His mission is to help navigate the chaos in a calm and approachable way.

Tackling the flood of crises *can* be overwhelming, but Climate Cat knows simple steps you can take to make a difference. So, whenever you see him, know that it's a chance to pause and reflect. In the face of all that's happening around us, every effort counts. And that's something Climate Cat wants you to remember.

BECOMING CLIMATE HEALTH LITERATE

"Climate change is the biggest global health threat of the 21st century."[1]

You wake up one morning to the sound of an alarm on your phone. It's an emergency alert: "Flash flood warning. Power outages are expected. Damaging winds. Prepare for road closures. Seek higher ground immediately." You panic as you realize you ignored the previous warnings. The floodwaters are rising, the power is out, and you're trapped.

We often brush aside these warnings, assuming they won't apply to us. We hear about hurricanes, wildfires, or heat waves devastating communities elsewhere, affecting someone else. Then, one day, it happens to us. Scorching temperatures, tornadoes, and floods have become part of our daily lives. The same alerts we once ignored are now urgent calls for action. In disasters, preparedness is the difference between crisis and safety.

Being prepared isn't just about storing emergency supplies or evacuating in time. It's understanding the basics of taking care of our health. *Health literacy* is a term that's widely used in public health and education.[2] It refers to knowing how to read, understand, and act on health information. Examples include following a doctor's advice, understanding a prescription label, and knowing when to seek medical care.

Just as health literacy helps us navigate medical advice and treatment, *climate literacy* involves understanding how climate change alters weather patterns and the conditions that affect our health. Rising temperatures intensify heat stress. Poor air quality aggravates chronic diseases. Extreme weather events cause injuries and evacuations and contribute to mental health struggles. If we cannot connect these dots, we may overlook the direct impact on our bodies, communities, and healthcare systems.

Climate change is a public health emergency, and we're all vulnerable. "Climate change is *everything* change. Because the climate is the fabric on which we weave our lives."[3] *Climate health literacy* is the ability to differentiate credible, science-based sources of information from those intended to confuse, mislead, or attack. It encourages us to seek out trusted experts, organizations, and resources that rely on scientific evidence.

The Impacts of Climate Change Are Not Equally Distributed

Older adults, low-income households, and people with chronic health conditions face especially high risks. In the United States, millions live with ailments such as diabetes, COPD, or asthma, and a significant share of the population is over 65 or living with a disability. Many rely on Medicaid or CHIP for coverage, while tens of millions still struggle below the poverty line. These intersecting vulnerabilities often leave people without the resources or care they need to cope with climate-related threats.

Communities of color face greater exposure to pollution and have less access to green spaces that help reduce heat, placing them at a significant disadvantage. They are also more likely to live in urban heat islands, heavily polluted environments, flood-prone areas, or communities with poorly maintained infrastructure. In addition, they often lack reliable access to quality sanitation and drinking water.[4]

How This Book Works

The Crisis Is Here will help you become climate health literate. Each chapter explores a specific area where climate intersects with health. We will examine how climate change affects our food supply, mental health, and even our pets, addressing its ripple effects on daily life. Whether it is understanding how extreme heat waves strain our bodies, staying informed through emergency alerts, or tackling air quality challenges, each section offers tools, insights, and actionable steps to help you and your community adapt and thrive in a changing world.

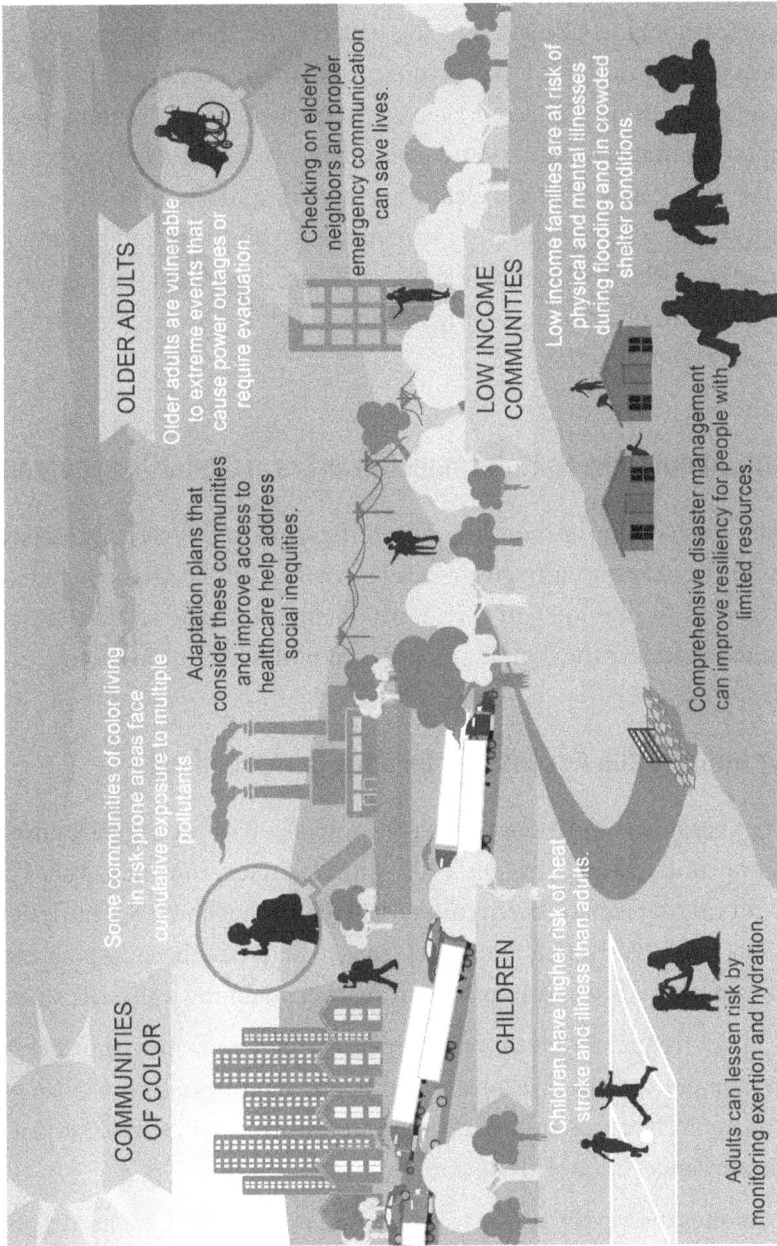

Source: U.S. EPA, via U.S. Global Change Research Program, "Chapter 14: Human Health," NCA4, 2018, https://web.archive.org/web/20250630045408/https://nca2018.global-change.gov/img/figure/figure14_2.png.

The book is designed to be modular. You do not need to read it cover to cover (though you're welcome to); each chapter stands alone, providing a deep dive into a single topic. If you're particularly concerned about wildfires, floods, or mental health, you can start there. If you want to prepare for power outages or understand *disinformation*, there's a chapter for that. By the end, you will have a clear understanding of the risks we face, the tools we can use, and the steps we can take to stay safe and healthy in a changing world. This is a story about resilience, hope, and finding ways to be healthy despite our constantly shifting world. To survive, we will adapt, act, and fight—together.

A Note About Links and Information in an Era of Digital Disappearance

I want to share my approach to providing links and resources throughout this book, especially given today's crisis of information availability. My goal is to provide you with trustworthy information that remains accessible despite the digital and political obstacles we now face.

Why Information Access Has Become Compromised

Since I began writing this book in the spring of 2024, the digital landscape has grown increasingly fragile and politicized. Government websites and scientific resources have been removed, altered, or shut down. Scientific information, particularly on topics like climate change, have become politicized. Waves of social media disinformation are overwhelming legitimate sources.

Links frequently break or disappear as content is moved or deleted. Government websites, research, and public data belong to us. We paid for them. When this information becomes inaccessible, it's not just an inconvenience, but a theft of something that's yours.

As sites disappear, are destroyed, or are manipulated, we need a physical book that preserves these links. While digital content is increasingly volatile, a book is a tangible archive. We don't know what will survive, so I am providing redundancy.

Here's how I've structured resources to help you navigate. For essential topics, I've deliberately included sources from different sectors: academic institutions and peer-reviewed journals, non-governmental organizations with established reputations, community-based resources with local expertise, and archive services that preserve deleted content. This is a strategy to ensure you can access crucial information even when specific sources become unavailable.

When Links Fail: Practical Recovery Strategies

If you encounter broken links or removed content:

1. Check Internet Archive's Wayback Machine (archive.org) for preserved versions.
2. Search for the same information across scientific journals, which maintain more stable archives.
3. Find alternative sources provided within the same section.
4. Try local libraries, which often maintain access to scientific databases.
5. Consider international sources that are not affected by U.S. policy changes.

Resources Over Recommendations

Rather than endorsing specific products or services, I share links to sites where you can make informed decisions. By providing multiple pathways to critical knowledge, political shifts or digital disruptions won't prevent you from accessing information.

JUST AS HEALTH LITERACY HELPS US NAVIGATE MEDICAL ADVICE AND TREATMENT, CLIMATE LITERACY INVOLVES UNDERSTANDING HOW CLIMATE CHANGE ALTERS WEATHER PATTERNS AND THE CONDITIONS THAT AFFECT OUR HEALTH.

CLIMATE CHANGE MAKES EVERYTHING WORSE

"Fire, heat, drought, and flood will transform our world this century."[1]

This epigraph is from one of my favorite books, *Nomad Century* by Gaia Vince. In it, Vince weaves history, science, and geopolitics to show that human migration has long been a survival strategy. One that will be intensified by climate change, as billions are displaced by extreme heat, rising seas, and resource scarcity. Vince argues that humanity's strength lies in our mobility, ingenuity, and ability to reshape societies to meet new challenges.

When discussing climate, four words are often used: "unprecedented," "extreme," "exacerbate," and "polycrisis."

Unprecedented reminds us that what was once rare is now the norm.

Extreme helps us see that these aren't just inconvenient events but life-altering catastrophes.

Exacerbate shows how one problem, such as a heat wave, can make other problems, like droughts or wildfires, significantly worse.

Polycrisis means we are not facing individual issues. Instead, we are dealing with multiple crises simultaneously, each one intensifying the others. A polycrisis occurs when climate disruption, economic and social instability, and health problems all collide.

In 2023, a series of devastating climate-related disasters underscored the escalating frequency and severity of extreme weather events:

⚡ **Canadian Wildfires:** Beginning in spring 2023, Canada experienced a record-setting wildfire season, with over 6,000 fires burning

37 million acres across all provinces and territories. This marked the largest area burned in Canada's recorded history. The fires resulted in the tragic loss of eight firefighters and displaced around 230,000 residents.

⚡ **Maui Wildfires:** In early August 2023, a series of wildfires, intensified by dry conditions and high winds, devastated the island of Maui, particularly the historic town of Lahaina. The fires resulted in at least 100 fatalities and destroyed over 2,200 structures, most of them residential. The estimated damages were $5.5 billion.

⚡ **Hurricane Idalia:** Hurricane Idalia made landfall in Florida as a Category 3 hurricane, bringing deadly storm surges and causing extensive damage across the southeastern United States. The hurricane led to widespread power outages and significant infrastructure damage. The estimated cost of damages ranged from $3 billion to $4 billion.

⚡ **Vermont Flooding:** In July 2023, historic flooding resulting from heavy rainfall submerged entire towns in Vermont. The floods caused extensive damage to homes, businesses, and infrastructure, leading to significant economic losses.

⚡ **Southwestern U.S. Heat Waves (June–August 2023):** The southwestern United States experienced record-breaking heat waves, with temperatures reaching unprecedented highs. The extreme heat led to increased hospitalization and fatalities due to heat-related illnesses and caused significant strain on power grids and water supplies. The combined drought and heat wave across the southern and midwestern U.S. from spring to fall of 2023 resulted in total costs of $14.5 billion.

A polycrisis is a web of interconnected challenges. It involves many problems occurring simultaneously and intensifying one another. These interrelated crises do not happen in isolation; each event feeds into the next, making every issue worse than it would have been on its own. A single domino falls and knocks over others, creating a chain reaction.

Climate Cat Explaining the Domino Effect

A *heat wave* doesn't mean more sweat and sunburns; it leads to heat exhaustion, dehydration, kidney failure, and death, particularly among vulnerable populations. These heat waves have become so extreme that they set the stage for catastrophic events, such as wildfires. Rising heat parches the land, turning forests, grasslands, and even cityscapes into tinderboxes.

Once a fire begins, it's often nearly impossible to contain, and the devastation spreads quickly, destroying homes, taking lives, and damaging ecosystems. When wildfires rage, they release massive amounts of carbon into the atmosphere, intensifying the global warming cycle. The added carbon raises temperatures further, leading to even more severe heat waves and fueling more fires. The dominoes continue to fall.

Wildfires add large amounts of CO_2 and soot to the atmosphere, which, over time, contribute to global warming and warmer oceans that can fuel stronger hurricanes. As hurricanes churn through, they bring wind, rain, and destruction, destabilizing entire communities. People already struggling with heat waves and wildfires must then face the added devastation of flooding, property damage, and displacement.

As temperatures rise, rainfall becomes more erratic, and water supplies become scarce. Droughts become more prolonged and intense, placing additional stress on agricultural systems. Crops fail, food production dips, and the price of basic goods soars. When people cannot access food or clean water, the result is social instability, political unrest, and climate migration.

As climate change worsens, the world will face increasingly interconnected disasters with a growing cost to human life. Every crisis feeds into the next, and every solution must address the complex web of challenges.

Here's how we can begin to understand these connections.

Hurricane Maria

Hurricane Maria's impact on Puerto Rico in 2017 was polycrisis at its worst. The entire island lost power, with some areas experiencing outages for months. The outdated and fragile electrical infrastructure was destroyed. Residents faced dire conditions without refrigeration for food and medicine or reliable access to clean water.

The healthcare system was *severely* damaged: hospitals were overwhelmed, supplies ran out, and many healthcare facilities were rendered inoperative. Conditions that were typically treatable became life-threatening as access to medicine grew increasingly scarce.

Hurricane Maria caused a near-total collapse of Puerto Rico's cellular network, with more than 88 percent of cell sites put out of service. The resulting communication blackout made it nearly impossible for residents to contact emergency services, reach loved ones, or receive critical updates.

THIS BOOK
IS ABOUT
MOVING FROM
REACTING TO
DISASTERS TO
PREPARING FOR
WHAT'S AHEAD.

The economy suffered a significant blow, with estimated damages exceeding $90 billion. Tourism, agriculture, and local businesses were hit particularly hard. Agricultural losses were immense: crops were destroyed, livestock lost, and food supplies and livelihoods severely affected. Many small businesses were wiped out, leading to widespread unemployment. Puerto Rico, a hub for pharmaceutical manufacturing, saw many facilities damaged, disrupting production and global supply chains for nearly a year.

The scale of climate change extends far beyond Puerto Rico and the United States, manifesting in devastating events worldwide. During Australia's 2019–2020 "Black Summer," bushfires burned about 24.3 million hectares, destroyed more than 3,000 homes, directly killed 34 people, and caused an estimated 417 additional deaths due to smoke exposure. The fires devastated wildlife, killing or displacing roughly three billion animals. The economic impact was profound, with estimated losses of 90 billion Australian dollars.

In July 2021, catastrophic flooding struck Western Europe, particularly Germany and Belgium. The disaster resulted in more than 243 deaths and widespread destruction. These floods were among the most severe in the region's recent history and underscored the escalating frequency and intensity of climate-induced disasters.

Europe's summer heat waves in 2022 caused more than 61,000 excess deaths. The same year, a drought labeled the worst in 500 years dried out about 60 percent of EU territory.

Extreme weather leaves lingering side effects that intertwine, producing consequences that reshape our landscape, communities, and public health. This overlap can be visualized as waves or converging forces that multiply the overall impact.

Impact of Climate Change on Human Health

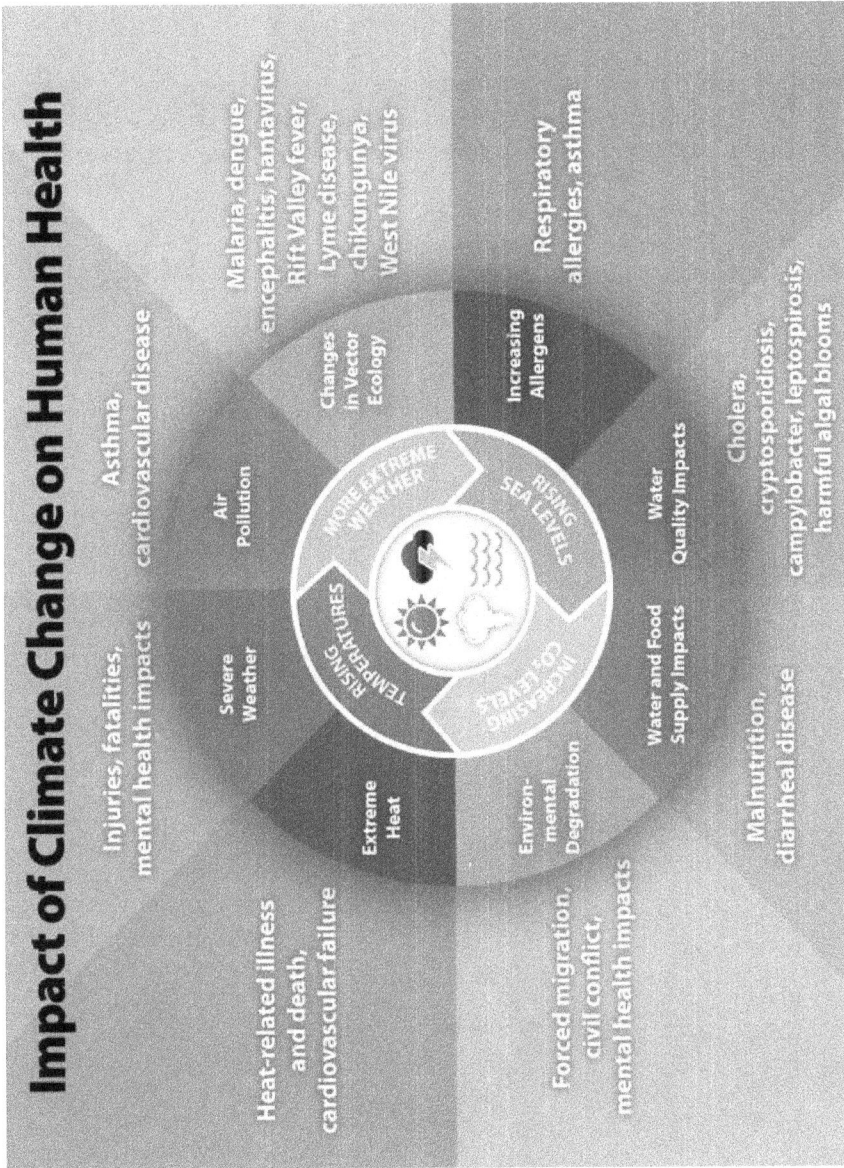

RISING TEMPERATURES

MORE EXTREME WEATHER

RISING SEA LEVELS

INCREASING CO₂ LEVELS

Extreme Heat — Heat-related illness and death, cardiovascular failure

Severe Weather — Injuries, fatalities, mental health impacts

Air Pollution — Asthma, cardiovascular disease

Changes in Vector Ecology — Malaria, dengue, encephalitis, hantavirus, Rift Valley fever, Lyme disease, chikungunya, West Nile virus

Increasing Allergens — Respiratory allergies, asthma

Water Quality Impacts — Cholera, cryptosporidiosis, campylobacter, leptospirosis, harmful algal blooms

Water and Food Supply Impacts — Malnutrition, diarrheal disease

Environmental Degradation — Forced migration, civil conflict, mental health impacts

Source: U.S. Department of Health and Human Services, HHS.gov, https://www.hhs.gov/sites/default/files/impact-of-climate-change-on-human-health.png.

Let's understand these impacts in more detail:

EXTREME WEATHER EVENT	IMMEDIATE HEALTH IMPACTS	WIDER CONSEQUENCES
Heat waves	Heatstroke Dehydration Cardiovascular and kidney stress	Power outages Water shortages
Wildfires	Smoke inhalation Burns Anxiety PTSD	Poor air quality Long-term lung damage
Floods	Drowning Water-borne illnesses Injuries	Disrupted medical care Mold exposure
Droughts	Malnutrition Dehydration Heat stress	Food insecurity Mental stress
Poor air quality	Asthma attacks Respiratory infections	Chronic lung disease Higher mortality
Vector-borne diseases	Lyme disease West Nile virus Allergic reactions	Greater risk for vulnerable groups
Food supply disruptions	Malnutrition Food poisoning	Supply-chain collapse Price spikes
Hurricanes and storms	Injuries Deaths Worsened chronic illness	Infrastructure damage Prolonged power outages
Water Contamination	Gastrointestinal illness Skin infections	Unsafe drinking water Crop failures
Displacement and overcrowding	Infectious disease spread Stress Trauma	Shelter crowding Exploitation risks
Mental health strain	Anxiety Depression PTSD	Long-term emotional and community strain

The crisis is here, reshaping our world and our health. Understanding these impacts is just the first step. This book is about moving from *reacting* to disasters to *preparing* for what's ahead. We can strengthen our communities, adapt our healthcare systems, and take action to minimize harm. The path forward requires resilience, innovation, and a commitment to change. But most importantly, it requires that we begin now.

EXTREME
WEATHER
EVENTS

CHAPTER 1

Heat and Health

"The climate impacts you hear about most often, from sea-level rise to drought to wildfires, are all second-order effects of a hotter planet. The first-order effect is heat." [1]

WHY IS IT GETTING SO HOT?

We are entering an era of extreme and unpredictable heat that will challenge our infrastructure, our health, and how we live. Events once considered anomalies are becoming commonplace. The 2003 European heat wave resulted in an estimated 70,000 deaths. Built for a temperate climate, homes and city infrastructure became death traps when temperatures soared beyond anything the region had previously endured.

Nearly two decades later, the Pacific Northwest faced a similar crisis when a staggering heat wave hit Seattle and Portland in 2021, shattering records and causing hundreds of deaths in a region unprepared for such extreme conditions. The heat will continue to rise year after year.

This isn't speculation; it's physics.

The planet absorbs more heat than it can release, and as greenhouse gases trap more energy in the atmosphere, temperatures continue to climb. Heat waves kill more people annually than hurricanes, floods, and tornadoes combined. They intensify droughts, fuel wildfires, and overload power grids, leading to blackouts when cooling is needed most.

Understanding the mechanics of this crisis is a survival skill. When the science of rising temperatures is understood, one can prepare, adapt, and advocate for change. Without that knowledge, communities stay vulnerable, and powerful interests exploit ignorance to delay action.

If you want to understand what's happening to your city, home, health, and future, you must understand heat. Climate change isn't just about melting icebergs; it affects the air you breathe, the safety of your neighborhood, and whether your family will have power during the next brutal summer.

Why Is It Getting So Hot?

Because the Earth functions like a living being.

In *Climate: A New Story*, Charles Eisenstein writes that the Earth is a living organism, and each ecosystem and species contributes to the health of the whole. Damage to any part of the planet impacts all life: "Each biome, local ecosystem, and species contributes in unique ways to the health and resiliency of the whole; they are the organs and tissues of the Gaian organism."[2]

Earth's life-support systems work in harmony to maintain stability and balance. Just as our bodies regulate temperature, pH levels, and hydration to stay healthy, Earth has its own mechanisms for regulating climate. These include the dynamic interplay of ice sheets, oceans, cloud formations, the atmosphere, and plant life.

Each system contributes to the planet's overall health and equilibrium. When one system fails, it disrupts the balance. This disruption is similar to how a fever or dehydration signals problems within our bodies that need immediate care.

The Greenhouse Effect

Gases like carbon dioxide (CO_2) and methane act like blankets, trapping heat in the Earth's atmosphere. CO_2 absorbs long-wave radiation emitted from the Earth's surface and reradiates it, warming the planet. The more CO_2 in the atmosphere, the more heat is retained, leading to rising global temperatures. Over the last forty years, temperatures have increased rapidly.

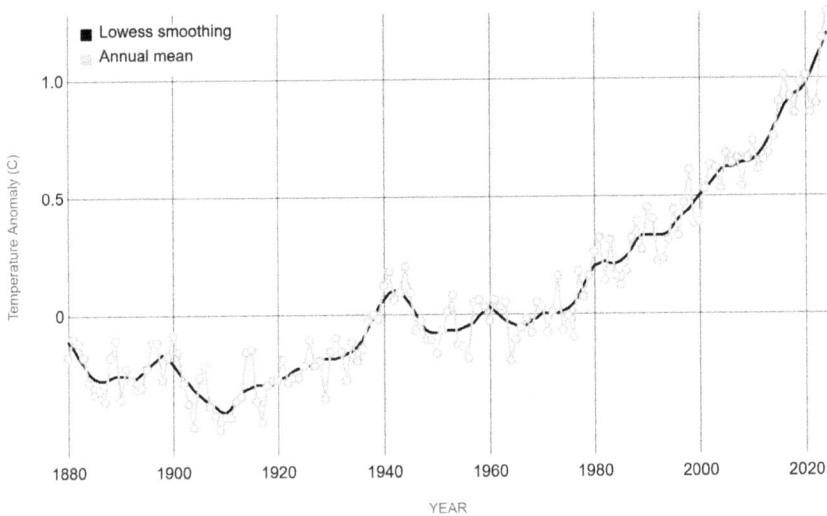

Source: NASA, "Global Temperature," NASA Climate Change: Vital Signs of the Planet, https://climate.nasa.gov/vital-signs/global-temperature/?intent=121.

Reflective Surfaces

Ice, snow, and clouds play a crucial role in regulating the Earth's temperature by reflecting a significant amount of solar radiation back into space, a process known as *the albedo effect*. As ice melts and snow cover diminishes due to warming, less solar energy is reflected. More heat is absorbed by the Earth's surface, further increasing temperatures in a feedback loop.

Oceans

Oceans distribute heat across the planet through currents like the Gulf Stream, moving warm water from the equator toward the poles. They act as carbon sinks, absorbing a large portion of the CO_2 from the atmosphere.

The ocean can only absorb a certain amount of carbon.

When the ocean absorbs CO_2, it reacts with seawater to form carbonic acid, lowering the ocean's pH in a process called ocean *acidification*. Acidification weakens the ocean's ability to absorb CO_2, meaning more carbon stays in the atmosphere, accelerating climate change. Acidification

3

makes it harder for marine organisms like corals, oysters, and other shell-fish to build and maintain their calcium carbonate structures, leading to weaker reefs, dissolving shells, and disrupted ecosystems.

Coral reefs, which support vast marine biodiversity and protect coastlines, are particularly vulnerable, with studies showing a significant decline in coral growth due to rising acidity. Shellfish industries have already suffered massive losses, as seen in U.S. oyster hatcheries, where larval oysters died when exposed to more acidic waters. Plankton (tiny organisms that form the base of the marine food chain) are also affected, which ripples up to larger fish and disrupts entire fisheries.

Plants

We're familiar with this aspect of CO_2 exchange, since we learned it in school. During photosynthesis, plants absorb CO_2 from the atmosphere, helping to lower atmospheric carbon levels and cool the planet. Deforestation and the degradation of ecosystems disrupt this balance. When forests are destroyed or degraded, they release stored carbon back into the atmosphere, turning natural carbon sinks into carbon sources and accelerating warming. Fires cause carbon *absorbers* to be carbon *emitters*.

Human Disruption of Earth's Systems

The climate crisis is not "natural."

It's the direct result of human activities that have disrupted Earth's delicate balance. Deforestation, soil erosion, ocean warming and acidification, habitat destruction, and the relentless release of greenhouse gases have destabilized the planet's climate. These disruptions drive extreme weather, rising sea levels, and ecosystem shifts, threatening food security, public health, and global stability. The more carbon we burn, the more we intensify these effects, pushing the climate system toward dangerous and unpredictable outcomes.

The impacts of warming are particularly severe in cities due to the *urban heat island (UHI) effect,* which can increase temperatures by up to 12 degrees. "Modern cities are empires of asphalt and concrete and steel, materials that absorb and amplify heat during the day, then radiate it out at night."[3] As of the 2020 Census, 80 percent of the U.S. population lives in urban areas, meaning most people will experience rising temperatures firsthand.

City infrastructure (buildings, roads, and concrete) absorb and retain heat, making urban areas much hotter than surrounding rural areas. Asphalt absorbs more sunlight and heats faster than natural surfaces like soil or grass. With fewer trees and green spaces to provide shade and cool the air through evapotranspiration, cities lack natural cooling mechanisms. Transportation, industry, and air conditioning units generate additional heat, raising temperatures. Urban areas have higher levels of air pollution, including greenhouse gases and particulate matter, which trap heat and worsen the overall warming trend.

What chain reactions are happening that will continue to increase heat?

Permafrost: Warmer temperatures cause permafrost to thaw, releasing trapped methane (a potent greenhouse gas) into the atmosphere, which amplifies warming.

Oceans: A warming atmosphere increases the heat stored in the oceans. Oceans act as massive heat reservoirs, and absorb about 90 percent of the excess heat from greenhouse gas emissions. Excess heat disrupts marine currents, like the Gulf Stream, which is critical in distributing heat across the globe. Changes in these currents lead to drastic regional temperature shifts, such as extreme heat waves in some areas and abnormal cold snaps in others.

Rain: When greenhouse gases trap heat, warmer temperatures cause more water to evaporate from oceans and land surfaces, increasing atmospheric moisture. Heightened moisture content contributes to more intense and frequent storms, heavy rainfall, and flooding in some regions, while others experience prolonged droughts.

WHAT TEMPERATURE IS IT?

How often have you looked at the temperature, and it *feels* hotter than the weather report? When checking the weather, you might notice "88°F" or "feels like 92°F." These numbers represent different ways of understanding temperature and how we experience it.

Let's go beyond the basic temperature reading and understand the components of measuring "heat."

Air Temperature (Dry Bulb Temperature): Air temperature measures how hot or cold the air is. It's the standard temperature reading you see on weather apps and thermometers. This measurement reflects the average kinetic energy of air molecules. When air molecules move faster, their kinetic energy increases, which we perceive as heat. While air temperature tells us the actual temperature, it doesn't account for how our bodies perceive it, especially when humidity is factored in.

The *Heat Index* combines air temperature and relative humidity to indicate how hot it feels to the human body. High humidity reduces the effectiveness of sweating, our body's primary cooling mechanism, making us feel hotter than the actual air temperature. For example, if the air temperature is 88°F with high humidity, the heat index might show a "feels like" temperature of 92°F or higher.

The heat index is a more accurate representation of the heat stress we experience. During warmer months, high humidity significantly increases the perceived temperature, heightening the risk of heat stress. The heat index is found in weather reports, apps, and health advisories. You can use it to plan outdoor activities, especially during peak heat hours.

Climate Cat Checks the Rising Heat

Understanding Heat Waves and Heat Alerts

A *heat wave* is a period of abnormally hot weather (and often high humidity) compared to the expected conditions of the area and the season.

Key characteristics of a heat wave:

Duration: Most heat waves typically last 3–7 days, though this can vary significantly by region. Climate scientists have observed that the average duration of heat waves has been increasing in many areas due to climate change.

Intensity: Temperatures significantly above the historical average for that location and time of year.

Impact: It can affect human health, power grids, transportation, and agriculture, and increase wildfire risk.

Regional Variation: What constitutes a heat wave varies by location (90°F might be a heat wave in Seattle but not in Phoenix).

Amplifying Factors: Urban heat islands can intensify heat waves in cities; high humidity makes heat more dangerous, reducing the body's ability to cool through sweating.

You can recognize an approaching heat wave in several ways:

Weather Alerts: Emergency notifications on your phone, weather apps, or local news.

Weather Forecasts: Check seven-day forecasts, especially in the summer months.

Physical Signs: Consistently rising temperatures over several days.

Official Announcements: From the National Weather Service or local authorities.

You Can Feel It: Unusually hot, still air, and a lack of cool overnight weather.

There are four main types of heat-related alerts on an increasing scale of severity:

1. **Extreme Heat Outlook:** A heads-up that very hot weather could arrive in the next 3 to 7 days. It's a signal to start paying attention.

2. **Extreme Heat Watch:** A warning that a dangerous heat wave is likely in the next 1 to 3 days. This is the time to get ready.

3. **Heat Advisory:** Very hot and humid weather is expected, usually with the heat index at 100°F or higher during the day and warm nights above 75°F. This type of heat can be dangerous, especially for kids, older adults, and people with health problems.

4. **Extreme Heat Warning** (the most serious alert): Extremely hot conditions are happening or about to happen. The heat index may hit 105°F or higher for days at a time, or even 115°F or higher for a shorter period. This is life-threatening heat. Without precautions, heat stroke, heat exhaustion, and other serious illnesses are very likely.

EXTREME HEAT WATCH

An Extreme Heat Watch is issued when **dangerous heat is *possible*.**

Reschedule outdoor activities in the coming days. Make sure that children, the elderly, and pets have a place to cool off during the heat.

Be Prepared.

EXTREME HEAT WARNING

An Extreme Heat Warning is issued when **dangerous heat is *happening or about to happen*.**

Avoid heavy activity & direct sunlight. Stay hydrated, find a cool indoor place, and check on children, elderly & pets.

Take Action!

weather.gov

Source: National Weather Service, Understanding Heat Alerts. https://www.weather.gov/images/safety/heat/heat-ww%20UPDATE%2029%20Jan%202025.png.

Wet Bulb Globe Temperature (WBGT)

"The wet-bulb temperature simulates a sweating human body, making it a better indicator of how well we can handle the air."[1]

Wet Bulb Globe Temperature is the most comprehensive measure of heat stress because it incorporates air temperature, humidity, solar radiation, and wind speed. It indicates how well the body can cool itself through sweating, making WBGT very useful for outdoor activities and high-heat environments.

Since WBGT measures heat stress in direct sunlight ("solar radiation"), it's more comprehensive than the heat index. You can find wet bulb globe temperature data at the following sources:

Zelus: https://zelusports.com/

Tempest Heat Alert: https://heatalert.app/

National Weather Service: https://www.weather.gov/ and https://digital.weather.gov/

Here's a simple way to use the different temperature readings:

MEASUREMENT	USEFUL FOR
Temperature	Use this basic measure to plan your day. If the air temperature is higher than usual, consider adjusting outdoor activities to other times, especially if you'll be in the sun.
Heat Index	Monitor the heat index closely during the summer. If it's high, stay hydrated, wear light clothing, and limit outdoor activities during peak heat hours.
WBGT	Use this for assessing safety during outdoor activities, exercise, and sports in hot environments.

High-Risk Populations for Extreme Heat

Extreme heat poses significant health risks, affecting people differently based on age, health status, and environmental factors. Understanding these risks is essential for effective protection.

DEMOGRAPHIC	HEALTH IMPACTS	ACTIONS
Children and Adolescents	Children have a lower heat tolerance due to their developing bodies and higher metabolic rates. Exercising in high temperatures is a leading cause of heat-related illness in this age group, with dehydration further impairing their ability to regulate temperature.	Ensure children are well hydrated before, during, and after activities. Schedule outdoor activities during cooler parts of the day. Provide frequent breaks in shaded or cool areas. Dress children in appropriate, breathable clothing.
Young Adults (Under 35)	Young adults under 35 are disproportionately affected by extreme heat, primarily due to occupational exposure and underestimation of heat risks. Studies have shown a significant rise in heat-related deaths among this group.	Take regular breaks in cool, shaded areas during outdoor work. Stay informed about weather forecasts and heat advisories. Educate the people around you about the signs and symptoms of heat-related illnesses.
Older Adults (Over 65)	Older adults have challenges with heat regulation due to age-related physiological changes. Frequent exposure to high temperatures accelerates biological aging, and chronic health conditions further compromise heat tolerance.	Stay in air-conditioned environments during peak heat. Increase fluid intake, especially water. Wear lightweight, loose-fitting, and light-colored clothing. Limit strenuous activities and rest frequently.
Pregnant and Breastfeeding Women	Pregnant and breastfeeding women are vulnerable to heat stress, which can lead to adverse outcomes like preterm birth, low birth weight, and stillbirth. Elevated temperatures are linked to increased risks of these complications.	Maintain adequate hydration, especially in hot weather. Avoid prolonged exposure to high temperatures. Wear loose, comfortable clothing. Consult healthcare providers for personalized advice.

DEMOGRAPHIC	HEALTH IMPACTS	ACTIONS
People with Chronic Health Conditions	Chronic health conditions, such as heart disease, diabetes, and respiratory issues impair the body's response to heat. Some medications hinder the body's natural cooling mechanisms.	Monitor health status regularly and adjust medications as advised. Stay in cool environments during extreme heat. Engage in only light physical activity while avoiding peak heat times. Maintain communication with healthcare providers.
Athletes and Outdoor Workers	Athletes and outdoor workers are at risk due to physical exertion and environmental conditions, which can increase heat-related illnesses.	Acclimate gradually to hot environments. Schedule activities during cooler times of the day. Ensure access to water and take regular hydration breaks. Wear appropriate, breathable clothing.

Synthetic Surfaces

Synthetic turf fields in public parks use crumb rubber infill made from recycled tires. These surfaces become much hotter than natural grass, posing health risks from heat. Elevated surface temperatures on synthetic turf can lead to heat-related illnesses, including heat exhaustion and heat stroke, especially during peak sun exposure. A youth soccer coach I know has said that artificial turf *has melted his athletes' cleats*.[2]

School Sports

In the past 32 years, at least 58 high school football players in the U.S. have died from heat stroke after collapsing on the field. Thousands more have experienced heat-related illnesses. High school football players are eleven times more likely to suffer from heat illnesses than all athletes in other sports combined.[3] In the summer of 2024, five high school football players died from suspected heat-related illnesses. Female high school cross-country athletes are twice as susceptible to heat-related illnesses than their peers in different sports.

EXTREME HEAT POSES SIGNIFICANT HEALTH RISKS, AFFECTING PEOPLE DIFFERENTLY BASED ON AGE, HEALTH STATUS, AND ENVIRONMENTAL FACTORS. UNDERSTANDING THESE RISKS IS ESSENTIAL FOR EFFECTIVE PROTECTION.

Schools monitor the heat index using weather forecasts, thermometers, and hygrometers (to measure humidity). Based on the heat index, they may set thresholds to modify, reschedule, move, or cancel outdoor activities to protect students. While guidelines can vary by district, many follow the general National Weather Service heat index categories:

Caution (80–90°F): Encourage frequent water breaks, reduce activity intensity, and provide shaded rest.

Extreme Caution (90–103°F): Limit outdoor activities, increase breaks, and closely monitor students for heat stress.

Danger (103–124°F): Move activities indoors or reschedule; outdoor sports should generally be suspended.

Extreme Danger (>125°F): Cancel all outdoor activities due to the risk of heat stroke and other severe health problems.

Pay attention to wet bulb globe temperature readings, particularly during heat waves. A reading above 95°F indicates hazardous conditions where the body cannot cool itself effectively. In such cases, prioritize cooling measures and avoid strenuous activities. However, studies suggest that even lower wet-bulb temperatures, such as 88°F, pose significant health risks, especially for vulnerable populations like seniors, children, and those with preexisting health conditions.[4]

The Workplace

The Occupational Safety and Health Administration (OSHA) sets and enforces workplace safety and health regulations. The National Institute for Occupational Safety and Health (NIOSH) conducts research and offers recommendations to prevent work-related injuries and illnesses.

The OSHA-NIOSH Heat Safety Tool App provides real-time heat index readings and first aid guidance for heat-related symptoms. Access it here: https://www.cdc.gov/niosh/heat-stress/communication-resources/app.html.

For workplaces, especially outdoor ones, OSHA guidelines provide specific actions based on heat index levels:

Less than 91°F: *Lower risk.* Basic heat safety practices apply: provide drinking water, allow regular rest breaks, and watch for signs of heat illness.

91–103°F: *Moderate risk.* Add precautions: schedule more frequent breaks, provide shaded or cool rest areas, and train supervisors to recognize early signs of heat illness.

103–115°F: *High risk.* Take extra protective measures: actively encourage hydration, adjust work schedules (earlier or later in the day), rotate workers, and closely monitor anyone with risk factors.

Greater than 115°F: *Very high to extreme risk.* Aggressive protective measures are essential: reschedule non-essential work, significantly reduce exposure time, use cooling equipment (vests, fans, misters), and provide medical monitoring.

HEALTH IMPACTS OF EXTREME HEAT

"Nothing speaks more directly to the global impact of planetary warming on human health than what happens to a person exposed to extreme environmental heat."[1]

Think about summer. Stepping onto the beach, the warm sand between your toes, the sun shimmering on the water. For generations, summer has been a time for vacations, barbecues, and lazy days outdoors. But as the world gets hotter, a trip to the beach isn't just about remembering sunscreen. It's about checking the heat index, packing extra water, and making sure no one in your group is at risk of heat exhaustion. We've moved from discomfort to survival.

Heat isn't an inconvenience; it's a killer.

During the 2022 FIFA World Cup in Qatar, extreme temperatures and poor working conditions led to the deaths of migrant workers building stadiums and infrastructure. Many collapsed from heat stress, their bodies unable to cool down in the relentless sun. The World Cup was rescheduled from its traditional June–July timeframe to November–December to avoid the extreme summer heat, which can exceed 122°F.

We know what unchecked heat can do.

We saw that in the summer of 2003, a European heat wave killed 70,000 people. In 2021, a record-breaking heat dome smothered the Pacific Northwest, causing temperatures to soar to levels never seen before. The effects were apocalyptic: the heat "cooked more than 1 billion sea creatures, many in their shells; fruit was cooked on the trees; crops and buildings were burned; and hundreds of people died."[2]

Heat waves are increasingly deadly, reaching into every part of life: oceans, agriculture, infrastructure, and human health. People in the U.S. die prematurely every year due to heat-related illnesses like heat stroke, kidney failure, and cardiovascular strain. "Premature," in this case, means preventable deaths. Lives were cut short because their bodies could not withstand the rising temperatures. Under high-warming scenarios, projections suggest that annual heat-related deaths in the U.S. could soar to 110,000.[3] Hospitalizations for heart disease, dehydration, and organ failure spike during heat waves.

Some of the damage is irreversible.

Rising temperatures have profound impacts on mental well-being. Elevated temperatures are associated with increased rates of depression, anxiety, aggression, and suicide. High ambient temperatures can lead to a rise in psychiatric hospital admissions. Exposure to extreme heat is linked to higher suicide rates, with projections suggesting that continued temperature increases could result in thousands of additional suicides in the coming decades. Furthermore, research has established a connection between hotter weather and increased aggression, leading to higher incidences of violence and crime.

What Heat Does to the Body

Heat is a powerful force that challenges the body's balance. When exposed to high temperatures, the body works hard to keep its core temperature stable, using mechanisms designed to cool us. Blood vessels near the skin widen, allowing heat to escape, while sweat glands produce moisture that evaporates, carrying heat away.

This process, while effective, comes at a cost.

Sweating depletes the body of water and essential salts, and as temperatures climb, the systems that regulate heat can become overwhelmed. Dehydration, heat exhaustion, and even heat stroke are risks when the body's cooling mechanisms are pushed too far.

Recognizing the symptoms of heat-related illnesses and knowing how to respond can be a matter of life and death. By understanding what to look for and having a clear plan, you can be ready to act when the heat wave arrives. This means you can remain focused and alert in soaring temperatures.

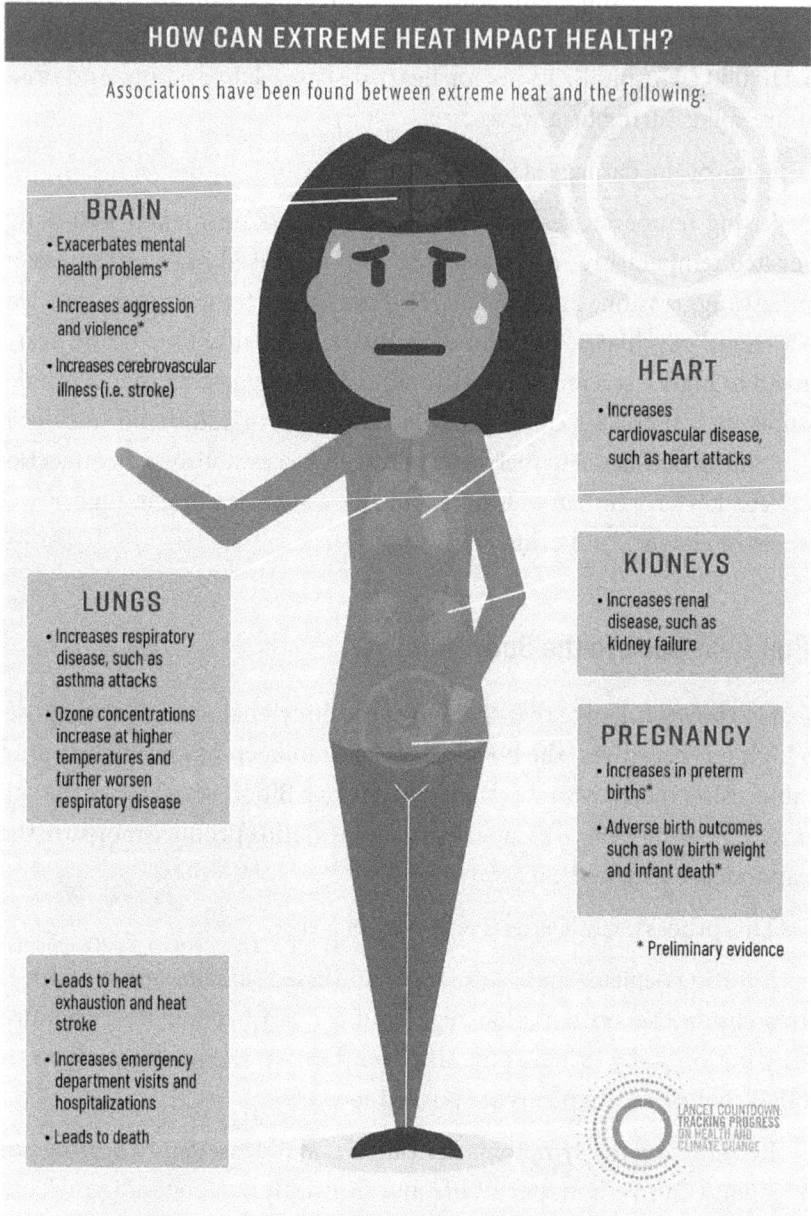

HOW CAN EXTREME HEAT IMPACT HEALTH?

Associations have been found between extreme heat and the following:

BRAIN
- Exacerbates mental health problems*
- Increases aggression and violence*
- Increases cerebrovascular illness (i.e. stroke)

HEART
- Increases cardiovascular disease, such as heart attacks

KIDNEYS
- Increases renal disease, such as kidney failure

LUNGS
- Increases respiratory disease, such as asthma attacks
- Ozone concentrations increase at higher temperatures and further worsen respiratory disease

PREGNANCY
- Increases in preterm births*
- Adverse birth outcomes such as low birth weight and infant death*

* Preliminary evidence

- Leads to heat exhaustion and heat stroke
- Increases emergency department visits and hospitalizations
- Leads to death

LANCET COUNTDOWN
TRACKING PROGRESS
ON HEALTH AND
CLIMATE CHANGE

Figure created for Brief by M. Lee (Climate Nexus).

Source: Lancet Countdown, 2018: 2018 Lancet Countdown on Health and Climate Change Brief for the United States of America. Salas RN, Knappenberger P, Hess JJ. Lancet Countdown U.S. Brief, London, United Kingdom, 32 pp.

The National Weather Service: Heat Index Tools page explains how to calculate the heat index, outlines the health risks of prolonged heat exposure, and offers guidance for preventing and managing heat-related illnesses. As heat waves become more common due to climate change, tracking the heat index should become part of your everyday routine.[4]

HEAT INDEX/HEAT DISORDERS

HEAT INDEX	POSSIBLE HEAT DISORDERS FOR PEOPLE IN HIGHER RISK GROUPS
80-90	Fatigue possible with prolonged exposure and/or physical activity.
90-105	Sunstroke, heat cramps and heat exhaustion possible with prolonged exposure and/or physical activity.
105-130	Sunstroke, heat cramps or heat exhaustion likely, and heat stroke possible with prolonged exposure and/or physical activity.
130 or higher	Heatstroke/sunstroke highly likely with continued exposure.

Source: National Weather Service, Heat Forecast Tools, Weather.gov, https://www.weather.gov/rnk/outdoorsheat.

Heat-related illnesses, like heat cramps, heat exhaustion, and heat stroke, happen quickly during periods of extreme heat and humidity. Early warning signs include muscle spasms (which may signal heat cramps), excessive sweating or dizziness (which can indicate heat exhaustion), and confusion or loss of consciousness (a critical symptom of heat stroke).

Know the Symptoms of Heat-Related Illnesses

SYMPTOMS

TREATMENT

HEAT CRAMPS

- Heavy sweating
- Painful muscle cramps or spasms

- Stop activity for a few hours.
- Move to a cooler location.
- Drink water, clear juice, or a sports beverage.
- Seek medical attention if cramps do not subside within one hour.

HEAT EXHAUSTION

- Heavy sweating
- Weakness
- Fatigue
- Headache
- Dizziness
- Nausea or vomiting
- Fainting
- Irritability
- Thirst
- Decreased urine output

- Move to an air-conditioned environment.
- Lie down.
- Loosen clothing or change into lightweight clothing.
- Sip cool, non-alcoholic beverages.
- Take a cool shower or bath, or apply cool, wet cloths to as much of the body as possible.
- Seek medical attention if symptoms worsen or last longer than one hour, or if the victim has heart problems or high blood pressure.

HEAT STROKE

- Very high body temperature
- Altered mental state
- Throbbing headache
- Confusion
- Nausea
- Dizziness
- Hot, dry skin or profuse sweating
- Unconsciousness

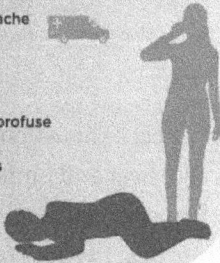

- Call 911 immediately and follow the operator's directions— this is a medical emergency.
- Reduce the person's body temperature with whatever methods you can: wrap the person in cool cloths, immerse them in a cool bath, or spray them with cool hose water.
- After administering cooling methods, move the person to a cooler place.
- Do NOT give liquids.
- If there is uncontrollable muscle twitching, keep the victim safe, but do not place any objects in his or her mouth.
- If there is vomiting, turn the victim on his or her side to keep the airway open.

Source: CDC, 2012

Source: U.S. Environmental Protection Agency, Know the Symptoms of Heat-Related Illnesses Infographic, https://www.epa.gov/emergencies-iaq/know-symptoms-heat-related-illnesses-infographic.

Other Heat Impacts

Cognitive Impairment: Confusion, impaired judgment, and slow reaction times. Actions:

- Hydrate
- Rest
- Cool the skin. Rest in wet clothing for faster cooling.

Respiratory Difficulties: Heat increases the production of ground-level ozone, which worsens asthma and chronic obstructive pulmonary disease (COPD), irritates the airways, and reduces lung function. Actions:

- Check the *air quality index (AQI)* and avoid outdoor activities on high-heat days.
- Use air purifiers to remove pollutants and allergens from indoor air.
- Use a dehumidifier set to 30 to 50 percent humidity. When indoor humidity is too low, it can irritate airways, making breathing difficult and aggravating respiratory conditions.
- Keep windows closed.
- Hydrate to keep the airways moist and reduce the risk of irritation.
- Wear a mask, especially in high-pollution areas or during pollen seasons.

Dehydration: Cramps, disorientation, and even cardiac arrhythmias (heart rhythm disorders). Actions:

- Drink 8 ounces of water every 15–20 minutes.
- Seek medical attention in cases of severe dehydration. IV fluids may be needed for rapid rehydration, especially if symptoms like dizziness, confusion, or fainting are present.

Be Alert for These Signs in Children

Signs in children to be alert for include increased thirst, weakness, irritability, headache, dizziness or fainting, muscle cramps, nausea and/or vomiting, increased sweating, cool and clammy skin, and a raised body temperature (but less than 104°F).[5]

Proactive Preparation

1. Stay informed about weather forecasts and heat advisories. The National Weather Service and weather apps offer heat risk forecasting tools.

2. Ensure older adults, young children, and people with chronic conditions remain safe and comfortable during high-heat events.

3. Talk about heat preparedness with your family and have a plan to stay cool and calm.

4. Drink plenty of fluids, and avoid alcohol and caffeine. Replenish electrolytes with sports drinks or foods rich in potassium and sodium. Keep a water bottle with you and drink regularly. Light yellow or clear urine usually indicates good hydration. If your urine is dark, drink more water.

5. A drop in weight from one day to the next, especially in hot weather, is often due to dehydration rather than fat loss. "Weight is your best indicator for dehydration: check your weight daily when it is hot."[6]

6. Use cooling strategies, such as air conditioning, fans, and cool showers. Avoid outdoor activities during the hottest parts of the day.

7. Locate cooling centers in your state. The National Center for Healthy Housing (NCHH) provides a state-by-state directory to help you find cooling centers near you. These centers offer air-conditioned spaces during extreme heat events and are especially critical for individuals without adequate cooling in their homes. Access here: https://nchh.org/information-and-evidence/learn-about-healthy-housing/emergencies/extreme-heat/cooling-centers-by-state/.

8. Recognize and respond to early signs of heat stress, like muscle cramps, dizziness, and fatigue.

9. Keep medications cool: heat can degrade medications, including antibiotics, aspirin, hydrocortisone, diabetic test strips, birth control pills, insulin, and seizure drugs. "Be extra careful with life-saving medications like EpiPens, nitroglycerin, glucagon, or naloxone. If these become damaged by heat, they may not be effective in an emergency."[7]

Understanding the UV Index

Our planet's natural defense, the ozone layer, protects us from harmful ultraviolet (UV) radiation. Thanks to the Montreal Protocol, the ban on ozone-destroying chemicals, the layer is slowly recovering. But it is still vulnerable.

Climate change and ozone depletion are linked. Rising greenhouse gases cool the stratosphere and shift wind patterns, which can affect how ozone forms and moves. In the Southern Hemisphere, especially over Antarctica, ozone thinning causes seasonal spikes in UV radiation. Recent research shows that massive wildfires, like the 2019–20 Australian fires, can inject smoke into the stratosphere and worsen ozone loss.

Closer to the ground, UV exposure increases at higher elevations because thinner air blocks less radiation. Snow and ice also reflect UV, which means skiers and mountaineers get extra exposure. With these factors combined, people everywhere, especially in mountain regions, benefit from taking precautions like wearing sunscreen, sunglasses, and protective clothing.

The UV Index measures ultraviolet radiation strength on a scale from 0 (low) to 11 (extreme). When the index reaches 8, unprotected skin can burn within just 10 minutes. You can check the daily UV Index through:

↯ Weather apps (Weather.com or AccuWeather.com)

↯ The National Weather Service website

↯ Google searches for "UV Index" plus your location

↯ Local news reports

Children's Skin Health Risks

While we often associate skin cancer with adults, children are also vulnerable. The three main types to be aware of are:

- **Basal Cell Carcinoma (BCC):** The least aggressive form, usually appearing on sun-exposed areas like the face.

- **Squamous Cell Carcinoma (SCC):** More likely to spread, appearing as scaly patches or sores.

- **Melanoma:** The most dangerous type (though rare in children), often presenting as changes in moles or new, irregular growths.

The ABCDE Rule for Monitoring Moles

Parents should regularly check their children's skin using the ABCDE rule:

- **Asymmetry:** One half doesn't match the other.
- **Border irregularity:** Edges are uneven or poorly defined.
- **Color variations:** Multiple shades of brown, black, or red.
- **Diameter:** Larger than 6mm (about the size of a pencil eraser).
- **Evolving:** Changes in size, shape, or color over time.

Excessive sun exposure can lead to serious health issues in children, including rapid onset of heat-related illnesses like heat exhaustion and heatstroke. Harmful UV rays can damage their eyes, raising the risk of cataracts and photokeratitis (sunburn of the eye), and prolonged exposure may weaken the immune system, leaving them more vulnerable to infections.

To protect your family from harmful sun exposure, apply broad-spectrum SPF 30+ sunscreen 15 to 30 minutes before going outside, and reapply it every two hours or after swimming. Dress children in sun-protective clothing, including long-sleeved shirts, wide-brimmed hats, and fabrics with a UPF rating. Avoid outdoor activities during peak UV hours between 10:00 a.m. and 4:00 p.m., and use umbrellas or stay in the shade whenever possible.

Keep children well hydrated, and include antioxidant-rich foods such as berries and omega-3 sources in their diet to support skin health and reduce inflammation. Technology can also support sun safety. Smartphone apps such as UVLens (https://www.uvlens.com/) provide personalized sun protection recommendations based on your location, skin type, and the current UV index.

Kidney Stones

Rising temperatures due to climate change are associated with increasing rates of kidney stones.[8] Higher temperatures lead to increased sweating, resulting in lower urine output and more concentrated urine, which promotes the formation of kidney stones.

Lemons are rich in citric acid, which inhibit the crystallization of calcium and other minerals, reducing the risk of stone formation. Drinking lemon water increases urinary citrate levels, creating a less favorable environment for developing kidney stones. In addition to lemon water, think about ways to increase your fluid intake. Drinking sufficient water dilutes urine, reducing the concentration of stone-forming substances.

How to Cool Off When the Power Is Out

When the power goes out during extreme heat, whether from a hurricane, tornado, or the strain of excessive temperatures overloading the grid, staying cool becomes a survival challenge.

Heat waves push the body to its limits, but the risks compound quickly when air conditioning, fans, and refrigerators stop working. Without power, dehydration sets in faster, food spoils, and medical devices that rely on electricity can become useless. To keep your home and body cool during extreme heat, try a variety of simple and effective methods.

✓ Keep cold packs or water bottles in the freezer to use for cooling beds or couches.

✓ For evaporative cooling, dip sheets in water and hang them in windows so that the breeze can blow through and cool the space. You can sleep on damp sheets for added relief.

✓ Close the doors to rooms that face the sun to prevent heat from spreading to cooler parts of the house, and spend time in the coolest rooms available, such as north-facing spaces or the basement.

✓ Use curtains, towels, or thick textiles such as duvets to block heat and keep hot air from seeping into the areas you want to stay cool.

✓ Cold water techniques can also be helpful. Take cold showers or baths, keep your hair wet, and apply a damp cloth or bandana to your neck, head, or wrists. For additional cooling, place a wet cloth under your armpits or on your groin, as these areas generate and retain the most heat.

✓ Keep battery-powered fans in your emergency kit to help circulate air, along with water-misting spray bottles, an ice chest for food and water, and a solar-powered battery for charging small electronics. You can place a pan of ice water in front of a fan to help cool the air.

✓ At night, open windows when the outside temperature is cooler, and close them in the morning to trap the cool air inside. Wait until the temperature drops, typically in the early morning or evening, before opening doors and windows again.

✓ Avoid cooking indoors if possible. Using the stove or oven can raise the temperature inside your home, so consider grilling or using a camping stove outside instead.

Potassium-Rich Foods

Extreme heat strains our bodies, increasing the risk of dehydration and heat-related illnesses. Potassium plays a critical role in managing these risks by helping maintain the body's balance of fluids and electrolytes, essential for proper muscle function, nerve signaling, and overall hydration. During intense heat, the body loses potassium through sweat, leading to fatigue, muscle weakness, and irregular heartbeats.

Potassium-rich foods help replenish these essential reserves, supporting the body's ability to regulate temperature, maintain energy levels, and prevent heat-related complications. When it's scorching outside, what you eat affects how your body handles the heat. Some foods help you stay cool, hydrated, and energized during the hottest parts of the day, while others provide essential nutrients that help your body recover and regulate temperature.

Best Foods to Eat in the Middle of a Hot Day

These foods and drinks are refreshing, hydrating, and easy to eat on the go. Oranges and orange juice are packed with potassium and vitamin C, making them both nutritious and hydrating.

- Peaches provide natural sugars for energy and are especially juicy and refreshing.
- Watermelon is made up mostly of water, offering a cooling effect along with a dose of potassium.
- Cantaloupe and honeydew melon support hydration and help maintain electrolyte balance.
- Kiwi and mango are tropical fruits that provide hydration along with vitamins and potassium.
- When fresh fruit is not available, dried fruits such as apricots, raisins, and dates serve as compact sources of potassium.
- Coconut water is another excellent option, as it naturally replenishes lost minerals and serves as a light electrolyte drink.

*Staying Cool Takes Planning,
Awareness, and Helping Others*

CLIMATE CAT'S HEAT TIPS

Refreshments: Steer clear of alcohol and caffeine, and opt for water or lemon water instead.

The Shadow Knows: Shade, cool rooms, and air conditioning.

Mind Your Meds: Certain medications can cause seniors to feel cold even in hot conditions and dull their sense of thirst, making dehydration more likely. Drugs such as diuretics, antihistamines, and antipsychotics can raise the risk of heat-related illnesses.

Clothes: Light-colored, loose-fitting, and lightweight.

Cool Cats: Apply a cool, wet cloth. Take a cool shower or bath. Mist the skin with water and allow it to evaporate.

Gastronomy Domine: Heavy meals raise your body's metabolic heat, fill your stomach, and can make dehydration worse. Stick to light meals instead.

Consideration for Others: Keep a vigilant eye on your *camerados* at the twin horizons of incarnation. The young who rush with the unbridled enthusiasm of kittens. Older people, who carry the wisdom of the ages, schooled on hours of *Jeopardy* and *Family Feud*. Both are uniquely susceptible to the sun's cruel and capricious rays.

CHAPTER 2

Air Quality

> *"What we call air is the record of everything we've burned."* —Climate Cat

WHAT ARE WE BREATHING?

Nitrogen and oxygen are not the only things we breathe in.

One of the most surprising aspects of writing this book was learning about the amount of garbage we inhale. "Air" includes all sorts of invisible particles: dust, pollen, chemicals, and pollutants from vehicles, factories, and wildfires. "They are so pernicious to human health that more than 7 million people die from air pollution each year."[1]

Breathing is the dance of life and chemistry within our bodies. With each breath in, oxygen-rich air travels down the trachea, branching into the bronchi. From there, it goes to the intricate network of bronchioles and then reaches the tiny, balloon-like alveoli in the lungs. Here, the thin walls of the alveoli and the capillaries meet, forming a respiratory membrane that oxygen diffuses across.

But the garbage gets processed in our bodies at the same time as the air.

Long-term exposure to air pollution is linked to chronic respiratory diseases, cardiovascular issues, lung cancer, and even neurological disorders like Parkinson's and Alzheimer's. If you have asthma or bronchitis, exposure to polluted air can trigger or worsen symptoms, leading to more frequent and severe attacks.

Before we dive into the air quality index and where all those invisible particles come from, let's understand the concept of particle size, called "PM," or *particulate matter*.

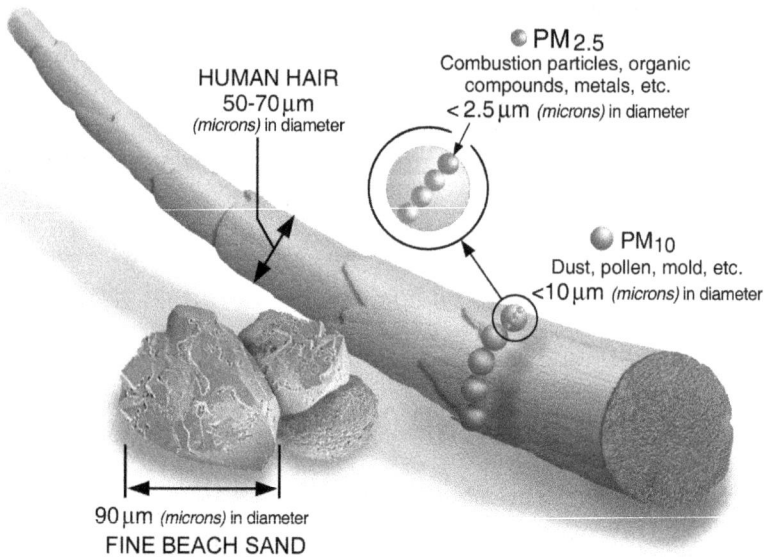

Source: U.S. Environmental Protection Agency, Particle Pollution (PM2.5 and PM10), EPA, https://www.epa.gov/sites/default/files/2016-09/pm2.5_scale_graphic-color_2.jpg.

The size of air pollutants plays a vital role in how they affect our health. Smaller particles, such as PM2.5, stay airborne for long periods and travel deeper into the respiratory tract, bypassing many of the body's natural defense mechanisms. Because they can reach sensitive areas of the lungs and even enter the bloodstream, these fine particles are linked to a wide range of health issues, including aggravated asthma, heart disease, and lung cancer.

The Air Quality Index (AQI)

Air quality is linked to climate change because many of the same pollutants, such as small particles and ground-level ozone, not only damage the air we breathe but also trap heat in the atmosphere. Over time, emissions from factories, vehicles, and deforestation have released more pollutants, worsening air quality in many cities. This, in turn, accelerates global warming, as higher temperatures lead to more wildfires that release additional pollutants.

"AIR" INCLUDES ALL SORTS OF INVISIBLE PARTICLES: DUST, POLLEN, CHEMICALS, AND POLLUTANTS FROM VEHICLES, FACTORIES, AND WILDFIRES. "THEY ARE SO PERNICIOUS TO HUMAN HEALTH THAT MORE THAN 7 MILLION PEOPLE DIE FROM AIR POLLUTION EACH YEAR."

The air quality index (AQI) is a daily rating that indicates how safe the air is to breathe, based on primary pollutants. The AQI measures how polluted the air is, or is forecast to become. The AQI typically includes ground-level ozone, particulate matter, carbon monoxide, sulfur dioxide, and nitrogen dioxide.

PM2.5 is widely considered the most harmful pollutant to human health. Ground-level ozone ranks high in toxicity, causing breathing difficulties and aggravating asthma. The other pollutants are also dangerous, but PM2.5 and ozone pose the greatest threat, especially in urban areas with heavy traffic or industrial activity.

The AQI measures the *concentration* of these pollutants and indicates potential risks on a scale of 0–500. Higher values indicate greater pollution levels and associated health risks:

UNDERSTANDING THE AIR QUALITY INDEX (AQI)

GOOD	AQI 0–50	No health impacts expected.
MODERATE	AQI 51–100	Very sensitive people may wish to limit outdoor exertion.
UNHEALTHY for sensitive groups	AQI 101–150	Sensitive people should limit outdoor exertion.
UNHEALTHY	AQI 151–200	Everyone should limit exertion outdoors.
VERY UNHEALTHY	AQI 201–300	Sensitive people should avoid all outdoor activity.
HAZARDOUS	AQI 301–500	Everyone should avoid all outdoor activity.

Source: National Weather Service, Understanding the air quality index (AQI) infographic, Weather-Ready Nation, Weather.gov, https://www.weather.gov/images/wrn/Infographics/2022/aqi-guide.png.

When AQI is above 100, "air quality is venturing into the unhealthy range—at first, for sensitive groups of people, then for everyone as AQI values increase."[2]

Understanding the Composition of the Air We Breathe

POLLUTANT	SOURCE
Nitrogen oxides (NOx)	Released from cars, trucks, airplanes, and power plants. They form smog and tiny particles that make it harder to breathe.
Carbon monoxide (CO)	A colorless, odorless gas emitted by cars, wood stoves, and fires. It reduces the oxygen in the bloodstream, which can be especially dangerous for people with heart conditions.
Volatile organic compounds (VOCs)	Emitted from vehicle exhaust, cleaning products, and paint.
Ozone (O_3)	A secondary pollutant created when VOCs and NOx mix with sunlight. It is a key component of smog and can cause breathing problems, with increased risk for people with asthma.
Photochemical smog	Often seen in large cities, it forms when sunlight reacts with pollutants like VOCs and NOx, creating a thick, brown haze.
Sulfuric acid (H_2SO_4) and nitric acid (HNO_3)	Acid rain develops when sulfur dioxide (SO_2) and NOx combine with water vapor, harming plants, buildings, and water sources.
Sulfurous smog	Sometimes called "London smog," it results from burning coal with high sulfur content, creating a thick, polluted fog.
Particulate matter (PM)	Tiny solid and liquid particles produced by diesel engines, construction sites, and wildfires. When inhaled, these particles can damage the lungs and heart.
PM10	Particles with diameters of 10 micrometers or less. Natural sources include dust, pollen, and mold spores; manufactured sources include construction sites and vehicle exhaust. PM10 can irritate the lungs and aggravate asthma.
PM2.5	Fine particles with diameters of 2.5 micrometers or less. They come from forest fires, vehicles, power plants, wood burning, and agriculture. These tiny particles can deeply penetrate the lungs and bloodstream, causing serious health issues.
Ultrafine particles (UFPs)	Particles smaller than 0.1 micrometers in diameter. They come from vehicle engines, industrial processes, wildfires, and even indoor cooking. UFPs can penetrate deep into the lungs, pass into the bloodstream, and reach the heart and brain, often carrying toxic chemicals.

Impacts on Children

Children breathe more air per unit of body weight than adults. Their immune systems are still developing, making it harder to fend off infections or inflammation caused by pollutants. If they spend more time outdoors, especially during periods of high air pollution, their exposure to harmful particles increases.

Children have a faster breathing rate than adults, which means they inhale more pollutants during the same period. Essential systems in children's bodies, such as the brain, lungs, and nervous system, are still developing. Air pollution can interfere with these developmental processes, leading to long-term health impacts. In 2018, the WHO reported that over 90 percent of children worldwide breathe toxic air daily, contributing to approximately 600,000 child deaths annually due to acute lower respiratory infections caused by polluted air.

Children are vulnerable to the neurotoxic effects of wildfire smoke. "Wildfire smoke is about 10 times as toxic as the regular air pollution we breathe from the burning of fossil fuels."[3] The fine particulate matter in wildfire smoke penetrates the respiratory system. It enters the bloodstream, reaching the brain and causing inflammation.

Children's nervous systems are still developing, so exposure to these pollutants has lasting impacts on their cognitive development and overall health. Parents of children with asthma face significant challenges during extreme weather, as wildfires and heat waves become more frequent. For instance, in British Columbia, one family lost their nine-year-old son to an asthma attack triggered by wildfire smoke. After his death, the family became advocates, working with the British Columbia Lung Foundation to raise awareness of the dangers of wildfire smoke and distribute air quality monitors to protect other children in their community.[4]

In Detroit, asthma is already a significant issue for children.[5] With rising temperatures and longer pollen seasons due to climate change, children experience more frequent asthma attacks. Parents rush their kids to emergency rooms when air quality worsens. Climate change and structural inequities, like living near highways with high traffic pollution, place even more strain on these families.

Other Groups at Risk

Seniors are highly vulnerable, as aging is accompanied by declining respiratory and cardiovascular health, making them more prone to diseases caused by air pollution.

Pregnant women are at risk, because exposure to air pollution is linked to adverse birth outcomes: low birth weight, preterm birth, and developmental delays. A systematic review published in *Environmental Health and Preventive Medicine* found consistent associations between maternal exposure to air pollutants like PM2.5 and adverse birth outcomes.[6]

People with preexisting health conditions, asthma, chronic obstructive pulmonary disease (COPD), heart disease, or diabetes are at risk. Asthmatics are more likely to experience aggravated symptoms, such as increased frequency of attacks, reduced lung function, and decreased quality of life due to exposure to pollutants.

Pollution impairs cognitive functions like attention span, memory, and decision-making.

A study published in *JAMA Psychiatry* found that long-term exposure to high levels of air pollution, particulate matter, and nitrogen dioxide (NO_2) is associated with an increased risk of depression and anxiety disorders. This study indicated that people living in areas with higher levels of air pollution had up to a 20 percent increased risk of developing depression and a 10 percent increased risk of anxiety.[7]

The Role of Wildfires

We've seen that wildfire smoke contains PM2.5-sized particles that lodge deep in our lungs, enter our bloodstream, cross the blood-brain barrier, and access our central nervous system. "Wildfire smoke" is high concentrations of these particles. Anyone, including healthy individuals, can be affected by wildfire smoke, even if the fire is hundreds of miles away.

Symptoms of wildfire smoke exposure include coughing, wheezing, and irritation of the eyes, throat, nose, and lungs. One of the most surreal examples was when New York City and much of the East Coast were shrouded in an eerie orange haze from the Canadian wildfires. The sky took on an apocalyptic glow, casting an unsettling filter over daily life that felt like something out of a dystopian movie.

Other Pollutants and Allergens

- ⚡ **Cancer-Causing Industrial Pollutants:** ProPublica created a detailed map of industrial pollutants across the U.S. Developed over two years, it analyzes billions of rows of EPA data to estimate the excess cancer risk at a neighborhood level. You can input your address to see localized information on potential carcinogen exposure. Access it here: https://projects.propublica.org/toxmap/.

- ⚡ **Dust Storms:** Plumes of Saharan dust travel over 5,000 miles from Africa to the Gulf Coast states, reducing air quality and affecting weather patterns. Dust particles can cause eye, nose, and throat irritation and reduce visibility, leading to car accidents and fatalities.

- ⚡ **Pollen:** Climate change is causing longer and more intense allergy seasons in the U.S. Warmer temperatures and higher carbon dioxide levels make plants produce more pollen for longer periods. Ragweed, a major allergen, now produces more pollen and stays active longer due to warmer falls and later frosts.

- ⚡ **Indoor Mold and Climate Change:** Warmer temperatures and increased humidity increase indoor moisture levels, promoting mold growth on walls, carpets, and furniture. Extreme weather events like hurricanes and floods create damp environments ideal for mold.

The Need for Stricter Air Quality Regulations: Ella's Law

Ella's Law, inspired by the death of nine-year-old Ella Kissi-Debrah in London, highlights the need for stringent air quality regulations. Ella, who had severe asthma exacerbated by air pollution, became the first person in the U.K. to have air pollution officially listed as a cause of death. Her case underscores the enormous health impacts of toxic air, particularly on children. It galvanized public and governmental action toward more aggressive policies to reduce air pollution. Clean air is not just a "privilege" but a fundamental right.

The next chapter will explore practical ways to protect your respiratory system from air pollution. Whether it's improving indoor air quality, staying informed about air quality levels, or learning about air filters, there are steps you can take to safeguard your health in a polluted world.

WHAT YOU CAN DO

When extreme weather, pollution, and wildfires create air quality emergencies across regions that once enjoyed clean air, each breath becomes a calculated risk. What was once as natural as breathing now requires careful monitoring and preparation.

"Studies have shown that most of a person's exposure to outdoor air pollution from wildfires happens while sheltering indoors."[1] Your home traps polluted air inside, especially when windows and doors are closed. Let's go over the steps to monitor air quality and protect yourself and your family when facing poor air conditions.

The first line of defense starts with awareness. Checking the air quality index (AQI) should become as routine as checking the weather. Here are tools that provide real-time updates on air pollution levels, helping you make informed decisions about outdoor activities, especially during wildfire smoke events or high-ozone days:

AirVisual: https://www.iqair.com/us/

AirNow: https://www.airnow.gov/

PurpleAir: https://map.purpleair.com/air-quality-standards-us-epa-aqi?

Create a safe indoor environment when outdoor air turns hostile. Think of your home as a clean-air fortress. Weather stripping and caulking become your first line of defense, sealing gaps that would let polluted air sneak in.[2] Your A/C system can switch from bringing in fresh air to recirculating what's already inside. The EPA's "Guide to Air Cleaners in the Home" provides a blueprint for transforming your living space: https://www.epa.gov/indoor-air-quality-iaq/guide-air-cleaners-home.

When selecting an *air purifier*, the Clean Air Delivery Rate (CADR) is a rating of how effectively it can protect your space. The higher the CADR, the better at cleaning the air.[3] Look for the basics: a good fan for circulation, a HEPA filter for catching tiny particles, and a washable pre-filter to extend your HEPA filter's lifespan.

CREATE A
SAFE INDOOR
ENVIRONMENT
WHEN OUTDOOR
AIR TURNS
HOSTILE.
THINK OF
YOUR HOME AS
A CLEAN-AIR
FORTRESS.

If you have an HVAC system, upgrading to a high-efficiency particulate air (HEPA) filter or a MERV-13 filter can transform your home into a clean air zone. MERV stands for Minimum Efficiency Reporting Value, the rating system used to measure the effectiveness of air filters in capturing particles of different sizes. A MERV-13 filter captures a significant percentage of airborne particles, making it suitable for improving indoor air quality.

> Don't have access to commercial filters? The EPA has instructions for building your own air purifier using a box fan and a MERV-rated furnace filter, a budget-friendly way to clean your air.[4]

Avoid activities that generate additional indoor air pollution, like vacuuming (which stirs up particles), burning candles, using gas stoves, smoking, or using aerosol products. If cooking is necessary, use exhaust fans that vent to the outside. According to the American Lung Association, avoiding these activities can reduce indoor particulate levels by up to 30 percent.

Keep children indoors during high AQI. When air quality is poor, stopping outdoor activities can prevent pollution exposure. Staying indoors helps reduce the risk of inhaling fine particles, which can penetrate deep into the lungs and even enter the bloodstream.

Limit outdoor time when the air quality index (AQI) is high, particularly during intense physical activities. The more strenuous your activity, the greater the risk. Plan outdoor activities when the AQI is lower, typically in the mornings. When being outdoors can't be avoided, use proper mask protection. For adults, N95 masks are best for adults, but children need special consideration: N95s aren't NIOSH-approved for younger kids.

Instead, kids over two can use properly sized KN95 or KF94 masks. For the tiniest family members under two, masks pose a suffocation risk, so consider specially designed stroller covers to shield them from harmful particles.[5]

Poor air quality doesn't just affect our lungs; it clouds our minds.

If you're already managing anxiety or depression, air quality emergencies can make these more intense. A few techniques to try:

✓ Diaphragmatic breathing engages your body's natural calming system. Imagine your belly as a balloon, slowly filling and emptying with each breath. You can inhale for a count of five seconds, hold for five seconds, and exhale for five.

✓ The 5-4-3-2-1 grounding technique is a simple way to bring yourself back to the present moment. Engaging your senses helps you break free from overwhelming emotions or thoughts and focus on your outer world. Start by identifying five things you can see around you, anything from objects in the room to trees outside. Next, tune into four things you can physically touch: the surface you're sitting on, the texture of your clothes, or a pet. Then, focus on three sounds you can hear: the person in the kitchen, music, or birds outside. After that, notice two smells you can detect. Finally, identify one thing you can taste: the lingering flavor of coffee, food in your mouth, mints, or gum.

✓ You can use emotional freedom techniques (EFT), the practice of tapping on specific acupressure points while focusing on releasing stress-inducing issues. Tapping can reduce anxiety and is associated with lower cortisol levels. Apps like the Tapping Solution guide you through the acupressure points. Access here: https://www.thetappingsolution.com/.

✓ The Climate Psychiatry Alliance offers guidance for families with children. They suggest turning breathing exercises into games during air quality emergencies. Try pretending to blow up an imaginary balloon, smell a flower, and blow out a candle. These playful approaches can help reduce anxiety while keeping kids engaged in healthy breathing practices.[6]

If you have asthma, COPD, or severe allergies, it's crucial to have a supply of essential medications, including nebulizers, allergy medications, and other prescription drugs. As we saw in the heat chapter, keep them in a cool, dry place so they remain effective.

Be alert for signs of cardiovascular or respiratory distress, such as shortness of breath, chest pain, wheezing, dizziness, and fatigue. The American Lung Association recommends working with healthcare providers to develop an "asthma action plan" or similar strategy for managing symptoms during wildfire events.

Protect Your Pets: Poor air quality puts your pets at risk. Limit their outdoor time during periods of poor air quality and create a safe indoor environment with clean air for them. Dogs and cats, like humans, are particularly vulnerable to respiratory problems caused by smoke and pollution. They have a limited ability to self-protect and require controlled outdoor exposure, along with special considerations for indoor air quality.

The American Lung Association's Lung HelpLine at 1-800-LUNGUSA is a free resource staffed by nurses and respiratory therapists who can answer questions about lung health, including protecting yourself during wildfires and periods of poor air quality. They offer practical advice for managing asthma, COPD, and other respiratory conditions when air quality is compromised.

Climate Cat Watches the Sky for Smog and Haze From Wildfires

CLIMATE CAT'S AIR QUALITY TIPS

Sniff the Breeze: Keys? Check. Phone? Check. Hold up. Peek at today's air quality index before bounding out the door. AQI in the red? That "fresh" morning walk can leave you panting or dizzy.

Fortify the Lair: Before wildfire smoke prowls or smog blankets the skyline, seal every crack with weather stripping or a dab of caulk. Switch your A/C to *recirculate*.

DIY Clean-Air Contraption: Strap a MERV-13 furnace filter to a humble box fan, and you've built a budget HEPA sidekick that scrubs the air like a diligent litter-box attendant.

Mask Like a Master: When the outside world feels like the inside of a chimney, armor up. N95s for grown-ups, KN95s or KF94s for kittens over two. Tinier paws (under two), opt for a stroller cover shield instead.

Soothe the Frazzled Feline Brain: Bad air isn't just a lung-lurker; it can claw at your mood. Try slow, deep "lion breaths," five-sense grounding, or playful bubble-breathing games to chase the anxiety away.

Pack the Preparedness Kit: Before the smoke alarm howls, stash spare filters, extra masks, critical meds, and emergency contacts in your "go bag" for grimy skies.

Protect the Tiny Tabbies: Little lungs gulp more air pound for pound. On smoky days, swap outdoor romps for indoor forts, and watch out for wheezes, coughs, or sudden energy crashes.

Know Your Territory: Homes near highways, factories, or wildfire corridors face stealthier pollutants. Add hotter temps and pollen bursts, and asthma/allergy flare-ups multiply.

CHAPTER 3
Natural Disasters

"One small flame can find a thousand miles of fuel." —Climate Cat

WILDFIRES

Wildfires have grown into a national crisis.

Once associated with the western states, fires now threaten areas that were once considered low-risk. Connecticut, where I live, experienced a fivefold increase in wildfires in 2024. The national statistics are alarming. In 2020, wildfires burned 10.3 million acres across the U.S. That figure is more than double the average in the 1980s. Wildfires now consume an average of 7 million acres annually, compared to just 3 million in the 1970s. What was once a seasonal threat has now become a year-round crisis.[1]

The frequency and intensity of wildfires continue to rise. Extreme wildfires are expected to increase by 14 percent by 2030, 30 percent by 2050, and 50 percent by 2100. Wildfires release vast amounts of carbon dioxide and other pollutants into the atmosphere, creating a feedback loop that increases the likelihood of future fires.

What Causes Wildfires?

Climate change, people, and poor fire management practices are key contributors.

The "ignition threshold" is when dry vegetation or forest debris reaches a temperature high enough to ignite and sustain a flame. Drier fuels, hotter conditions, and higher wind speeds cause fires to start and spread. Lightning, human activities, and embers from existing fires also trigger ignition.

Global temperatures are rising and have created a hotter, drier world that transforms forests and grasslands into tinderboxes. In 2024, Connecticut experienced its driest September and October in 120 years, with an eight-inch rainfall deficit, ideal conditions for wildfires.

Higher temperatures dry vegetation, reducing moisture levels in forests, grasslands, and brush. Reduced rainfall and changing weather patterns result in drier soils and extended dry seasons, making whole ecosystems susceptible to large-scale, uncontrollable wildfires.

Warmer temperatures generate thunderstorms and lightning, primary natural ignition sources for wildfires. A 2020 study found lightning strikes have increased by 20 percent in the western U.S. due to climate change.

Eighty-five percent of wildfires result from human actions, including unattended campfires, discarded cigarettes, and electrical malfunctions. As urban areas expand into wilderness, the risk of human-caused ignitions increases.

Decades of aggressive fire suppression have prevented many naturally occurring, low-intensity fires that clear away underbrush and excess vegetation. As a result, forests accumulate large amounts of dead trees, fallen branches, and dense undergrowth. This makes them prone to larger, hotter, and more destructive fires that are harder to control and pose greater risks to people, wildlife, and property.

Health Impacts

We've seen the health risks from poor air quality, but wildfires create unique and concentrated hazards beyond the flames. Studies show a rise in respiratory and cardiovascular hospital visits during heavy smoke events, underscoring the dangers for people with heart disease. The mental health toll is serious, with prolonged smoke exposure raising the risks of anxiety, depression, and even post-traumatic stress disorder (PTSD). Children often suffer the most, and the impacts can linger for weeks after the fires have been put out.

Health Risks of Wildfire Smoke Exposure

Respiratory Effects	Exposure to wildfire smoke can irritate the eyes, nose, and throat. Persistent coughing, and shortness of breath. It can aggravate asthma and COPD, increasing the risk of bronchitis and pneumonia.
Cardiovascular Impacts	Fine particulate matter from wildfire smoke raises inflammation levels and damages blood vessels, leading to heart attacks, strokes, and other cardiovascular issues.
Neurological Effects	Emerging research suggests that wildfire smoke exposure impairs cognitive function and increases the risk of neurodegenerative diseases.
Perinatal and Developmental Effects	Pregnant women exposed to wildfire smoke face higher risks of adverse birth outcomes, including preterm birth and low birth weight, which can affect child development.
Other Health Impacts	Wildfire smoke exposure is linked to inflammatory skin diseases like acne and may elevate the risk of certain cancers.

Preparing for Evacuation

Create an emergency plan:

1. Begin discussing evacuation planning, evacuation routes, and meeting points for your household. Include plans for your pets.

2. What are the different ways you could get out of your house in an emergency? (Doors, windows, basement, garage, stairs, ladder.)

3. The Red Cross recommends identifying at least *two* evacuation routes from your neighborhood and designating meeting locations: one close to home and another outside your neighborhood.

4. Practice these routes with all family members, including in low-visibility conditions that might occur during a fire.

Prepare an emergency kit:

- FEMA recommends storing items in easy-to-carry containers like plastic bins or duffel bags placed in an accessible location. For a discussion on emergency kits, including essential items, see Chapter "Emergency Kits."

Stay informed. Sign up for emergency alerts through:

- Local agencies; many use Everbridge Nixle (https://www.everbridge.com/products/nixle/)
- InciWeb: U.S. Forest Service Wildfire Information (https://inciweb.wildfire.gov/)
- FEMA App: Emergency Alerts and Resources (https://www.fema.gov/about/news-multimedia/mobile-products)

According to the National Weather Service, having multiple ways to receive warnings is crucial. Consider a battery-powered weather radio as a backup when power and cell service fail. Follow local fire departments and emergency services on social media platforms for real-time updates.

If time allows before evacuation, take steps to improve your home's chances of surviving a wildfire.

Move flammable outdoor furniture inside and remove combustible items, such as wood piles and propane tanks, from around the house. Close all windows, vents, and doors to prevent embers from entering and igniting indoor materials. Turning off the gas at the meter and extinguishing pilot lights can help reduce the risk of explosions or fire spread. Leave house lights on so firefighters can easily locate and navigate your home in smoky conditions. Filling sinks, tubs, and buckets with water can aid firefighting efforts.

When to Evacuate

Wildfires can escalate with alarming speed. Under hot, dry, windy conditions, the fire front may advance up to 14 miles per hour, and wind-blown embers can ignite new fires miles ahead. In such conditions, early evacuation is critical.

Waiting until smoke or fire is visible can leave too little time to escape safely.

Fires often move so quickly that roads become impassable, cutting off escape routes. The California Department of Forestry and Fire Protection (CAL FIRE) emphasizes that "when in doubt, get out." Waiting until the last minute puts you and emergency responders at risk.

Consider evacuating before an official order if:

1. You or family members have respiratory conditions, mobility issues, or other health concerns

2. Air quality is rapidly deteriorating (AQI above 200)

3. According to the U.S. Forest Service, these conditions warrant immediate evacuation, even without official orders:

   a. Rapidly increasing smoke or ash

   b. Visible flames approaching your location

   c. Strong, erratic winds in fire-prone areas

   d. Sudden temperature increases

   e. Unusual animal movement (wildlife fleeing the area)

   f. You hear sounds of approaching fire (popping, crackling)

   g. Roads to your area are beginning to close or become congested

   h. You feel unsafe for any reason

Special Considerations for Vulnerable Populations

The CDC advises that certain groups should consider evacuating earlier, as they may require additional time to travel safely and securely. These include seniors, children under five, pregnant women, people with respiratory or cardiovascular conditions, and people with mobility limitations.

WILDFIRES CAN
ESCALATE WITH
ALARMING SPEED.
UNDER HOT, DRY,
WINDY CONDITIONS,
THE FIRE FRONT
MAY ADVANCE UP
TO 14 MILES PER
HOUR, AND WIND-
BLOWN EMBERS CAN
IGNITE NEW FIRES
MILES AHEAD. IN
SUCH CONDITIONS,
EARLY EVACUATION
IS CRITICAL.

Official Evacuation Orders

Always follow mandatory evacuation orders issued by fire departments, emergency management agencies, or local government officials. These orders use a three-level system in many jurisdictions:

1. **Evacuation Warning/Alert:** Prepare to evacuate; gather essentials and be ready to leave.

2. **Evacuation Order:** Leave immediately; delaying puts your life at risk.

3. **Shelter in Place:** It's too dangerous to evacuate; stay where you are until conditions change.

The National Highway Traffic Safety Administration advises checking traffic apps for congestion along evacuation routes. Listen for radio reports about road closures. Have paper maps as backups for GPS navigation. Avoid routes that pass through canyons or heavily forested areas if possible. Keep at least half a tank of gas in your vehicle at all times during fire season.

After a Wildfire

Returning Home Safely

Return home only when authorities confirm it's safe, and stay alert for potential hazards.

Smoldering debris and unstable structures, such as weakened trees, power poles, and buildings, can pose serious risks. Be mindful of hot spots that may reignite, ash, and toxic residue that could contain hazardous chemicals.

Damaged utility lines, including downed power lines and potential gas leaks, require extra vigilance.

Displaced wildlife may have moved into the area, creating unexpected dangers.

The EPA warns that wildfire smoke particles can linger in homes for weeks after a fire is extinguished, continuing to pose respiratory risks. Wait until the air quality improves before returning, and be ready to wear an N95 mask during cleanup to protect yourself from lingering pollutants.

Proper cleanup after a wildfire protects your health and prevents further exposure to harmful particles.

Wear protective clothing, including long sleeves, pants, sturdy shoes, and work gloves, to minimize skin contact with ash and debris. Always use an N95 mask during cleanup to reduce inhalation of fine particles.

Before sweeping, wet down the ash to prevent it from becoming airborne. Avoid using regular vacuums, which can recirculate harmful dust; instead, use a HEPA vacuum. Dispose of ash properly by sealing it in plastic bags according to local regulations. Clean all surfaces with damp cloths rather than dry dusting to avoid stirring up particles.

The California Department of Public Health warns against using leaf blowers or standard vacuums for ash cleanup, as these can spread toxic particles throughout your home. Consider hiring professional cleaning services equipped with specialized tools to address extensive damage, ensuring a thorough and safe restoration process.

Discard any food that has been exposed to heat, smoke, or fire-retardant chemicals. The Red Cross advises throwing away any food not stored in sealed, undamaged containers and disposing of refrigerated items if the power is out for more than four hours. Food with an unusual odor, color, or texture should be thrown away.

If water contamination is suspected, use bottled water or boil tap water until authorities confirm it is safe to drink. Thoroughly clean all cookware and eating utensils before use, and wash homegrown fruits and vegetables carefully to remove any lingering contaminants.

Monitor your health closely after exposure to a wildfire, as symptoms of smoke inhalation may not appear until days later.

The CDC warns that certain signs require medical attention, including persistent coughing, wheezing, shortness of breath, chest pain, or heart palpitations. Unusual fatigue, weakness, or recurring headaches may indicate underlying issues caused by smoke exposure. Changes in mental state, confusion, or dizziness may signal serious health concerns. Prolonged eye irritation that does not improve could be a sign of continued exposure to airborne particles.

Seeking medical care for these symptoms can help prevent complications and ensure a full recovery. Keep a log of symptoms, particularly for vulnerable family members. The American Lung Association recommends consulting healthcare providers about preventative measures *before* returning to areas with lingering air quality issues.

Post-wildfire mental health concerns are common and should not be overlooked, according to the Climate Psychiatry Alliance. The emotional toll of a wildfire can manifest in depression, anxiety, or PTSD, so it's important to monitor yourself and your family members for these signs.

Connecting with wildfire survivors' community support groups can provide a sense of solidarity. Seeking professional mental health services can help process trauma and build coping strategies. Establishing routines can be beneficial, particularly for children, as it helps restore stability in the face of disruption. Prioritizing self-care and stress-reduction techniques, like mindfulness or physical activity, can aid in emotional recovery. Emotional healing often takes longer than physical rebuilding, and giving yourself grace during this process is essential.

Wildfire Health and Safety Resources

- **American Lung Association: Wildfire Smoke and Your Health**

  https://www.lung.org/clean-air/emergencies-and-natural-disasters/wildfires

  Provides information on how wildfire smoke affects lung health and offers tips to protect yourself during wildfire events.

- **Centers for Disease Control and Prevention (CDC): Wildfire Smoke and Public Health**

  https://www.cdc.gov/wildfires/risk-factors/index.html

  Explains the health effects of wildfire smoke exposure and offers guidance on protecting vulnerable populations.

- **California Department of Public Health (CDPH): Safe Cleanup of Fire Ash**

  https://www.cdph.ca.gov/Programs/EPO/Pages/Wildfire%20Pages/Safe-Cleanup-of-Ash.aspx

  Offers guidelines for safely cleaning up ash after a wildfire to minimize health risks.

- **Climate Psychiatry Alliance: Wildfire Smoke Toolkit**

  https://www.climatepsychiatry.org/wildfire-smoke-toolkit

  Provides resources on the mental health impacts of wildfire smoke and strategies to cope during air quality emergencies.

- **National Interagency Fire Center: Wildfire Statistics**

  https://www.nifc.gov/fire-information/statistics

  Offers comprehensive statistics on wildfires, including annual reports and historical data.

- **Ready.gov: Wildfires (U.S. Department of Homeland Security)**

  https://www.ready.gov/wildfires

  Provides information on how to prepare for, stay safe during, and recover from wildfires.

- **U.S. Forest Service: Wildfire Risk Assessment**

  https://wildfirerisk.org/

  Offers tools and resources to help communities understand and reduce their wildfire risk.

*Climate Cat Cools the Blaze
Before the Sparks Fly*

CLIMATE CAT'S FIRE SAFETY TIPS

Lock Down the Lair: When the outside air looks like burnt toast, don't invite extra grime indoors. Pause the vacuum's roar, skip the scented candles, and turn off the gas stove.

Choreograph Your Outdoor Cha-Cha: Sync your prowls with nature's cleanest beats, early dawn or after a cleansing rain. That's when the sky does its own house-cleaning, giving you fresher gulps of air for your outdoor adventures.

Sip the Elixir: Water is your body's silent janitor, flushing smoke toxins and keeping airways slick. Keep a bottle within paw's reach and lap it often. Hydrated cats (and humans) breathe easier.

Pack the Nine-Lives Kit: Stash N95 masks, everyday meds, bottled water, snacks, vital papers, and a battery radio in one grab-and-go tote. Be ready. Throw in extra kibble, a leash, and a carrier so the furriest family members evacuate.

Exit Before the Sirens: Got asthma, heart trouble, or slower paws in the household? Beat the rush and roll out before officials say "Go!" Early escapes cut stress and dodge gridlock.

Read Your Own Whiskers: Bodies whisper before they yowl. A nagging cough, chest tightness, dizziness, or labored breathing means "call the doctor or vet!" Kids and elders react fastest, so watch for odd fatigue, crankiness, or quick breathing.

Guard the Sensitive Pride: Tiny tabbies and wise lions need extra vigilance; their lungs are more delicate. Keep indoor games ready and monitor for wheezes or sudden slumps.

Stay in the Loop: Subscribe to local AQI alerts and emergency texts. A single ping can spare you a throat full of ash. Knowledge is a super-powerful catnip.

WATER

Water is life. Water can shape and reshape landscapes, destroying or sustaining us depending on how we respond to its power.

Between 2013 and 2022, flooding caused 57 percent of U.S. fatalities from tropical cyclones. "Floods are the most common disaster in the United States. Failing to evacuate flooded areas or entering flood waters can lead to injury or death."[1]

If flooding is the most common disaster in the U.S., it's also one of the most expensive ones. Hurricane Helene struck North Carolina in late September 2024, causing catastrophic flooding and landslides. The North Carolina Office of State Budget and Management estimated the total damage and recovery cost $60 billion. That's from just one hurricane.

What's a Flood?

A *flood* is an overflow of water that submerges land that's usually dry. Simple, to the point. The problem is that there are more and more of them. Here is a breakdown of different types of floods, when they occur, and what triggers them.

Understanding your flood zone is crucial for preparing for potential disasters. Flood zones, as mapped by FEMA, indicate areas at high risk for flooding, whether from hurricanes, heavy rain, or storm surges. Living in a designated flood zone requires flood insurance, elevated home foundations, and a clear evacuation plan.

TYPE OF FLOOD	CAUSE	FREQUENCY
Flash Floods	Heavy rainfall, dam failures, hurricanes	This can happen multiple times a year, especially in flood-prone areas.
Coastal Floods	Storm surges, rising tides, hurricanes	Frequent along coastlines, worsened by sea-level rise.
River Floods	Prolonged rainfall, rapid snowmelt	It can be seasonal, especially after heavy precipitation or rapid thaw.
Urban Floods	Overwhelmed drainage systems	Due to paved surfaces and extreme rain events.

Hurricanes and Flooding

Hurricanes are powerful, rotating storms that bring intense rainfall and strong winds.

Remember that heat is a form of energy, so warmer temperatures lead to stronger systems with heavier rainfall. Slower-moving hurricanes prolong the rain, increasing the risk of flooding. Rising ocean temperatures are driving hurricane activity farther north, affecting a wider range of regions in the country.

The North Atlantic has experienced an increase in hurricane intensity, frequency, and duration, with a higher proportion of Category 4 and 5 storms forming in recent decades. On average, hurricanes can bring 10–15 inches of rain, but extreme events, such as Hurricane Harvey in 2017, shattered records with 60 inches of rainfall in Texas. These heavy rains caused record-breaking floods and resulted in billions of dollars in recovery costs.

When hurricanes strike, they submerge entire neighborhoods. Streets turn into rivers, homes are left partially underwater, and communities are forced to evacuate. In some cases, only rooftops remain visible above the floodwaters. Vehicles are swept away, and contaminated water spreads, increasing health risks. The aftermath is a scene of devastation: crushed homes, scattered debris, and families struggling to recover.

Floodwaters do more than destroy houses and roads; they contaminate water supplies. As floodwaters surge, they pick up sewage, industrial waste, agricultural runoff, and harmful bacteria, polluting drinking water sources. This contamination causes outbreaks of waterborne diseases. There is a 35 percent increase in hospitalizations for infectious diseases, mostly from contaminated water, in the months following major hurricanes.

Flood runoff increases the presence of toxic chemicals and heavy metals in rivers and lakes. When floodwaters recede, they leave behind standing water that breeds mosquitoes, spreading diseases like West Nile virus. The connection between flooding and water quality is clear. Extreme rainfall and rising sea levels can contaminate water supplies, making it crucial to plan ahead for safe drinking water after disasters.

WHEN HURRICANES
STRIKE, THEY
SUBMERGE ENTIRE
NEIGHBORHOODS.
STREETS TURN INTO
RIVERS, HOMES
ARE LEFT PARTIALLY
UNDERWATER, AND
COMMUNITIES
ARE FORCED TO
EVACUATE. IN
SOME CASES, ONLY
ROOFTOPS REMAIN
VISIBLE ABOVE THE
FLOODWATERS.

Health Impacts of Hurricanes

Seniors account for 60 percent of hurricane-related casualties.

Evacuating can be challenging for people with limited mobility, and many older adults depend on medications

Forty percent of deaths during Hurricane Katrina were caused by drowning, and 25 percent by injury and trauma. Carbon monoxide poisoning (from improper generator use) also emerged as a significant cause of death.

that may be difficult to get during or after a disaster. Power outages during hurricanes increase the risk of heat stress, which can be particularly hazardous for older individuals who are more susceptible to extreme temperatures.

Low-income communities face multiple barriers during hurricanes, including difficulties with evacuation due to transportation and cost. These communities are often located in flood-prone areas, making them more susceptible to the devastation caused by rising waters. They may have limited insurance coverage, leaving them financially vulnerable when recovery is needed. These neighborhoods experience higher flood damage compared to wealthier areas, and recovery efforts take twice as long due to limited resources and longer wait times for assistance.

People who use medical devices are especially vulnerable during hurricanes.

If you rely on powered medical equipment, such as oxygen or ventilators, a power outage can jeopardize your health. Storage and transport of medications can become a problem if pharmacies are closed or supplies run out. Dialysis access can be interrupted, and emergency services may be difficult to reach, creating additional challenges if immediate medical care is needed.

Children face heightened risks as well. They are more susceptible to trauma and post-traumatic stress disorder (PTSD) due to the chaotic and dangerous circumstances of flooding. Waterborne illnesses can spread quickly, and children are more vulnerable to infection. Education is disrupted when schools are closed for extended periods, which can hinder both academic progress and emotional development.

What Is "Water Quality"?

Water quality measures the safety of water for drinking, swimming, and aquatic life. It measures bacteria, heavy metals, chemicals, and natural factors, such as pH and mineral content. To check your local water quality, you can look at your water company's annual water quality report (sometimes called a Consumer Confidence Report).

You can use the Environmental Working Group's Tap Water Database (https://www.ewg.org/tapwater/), or check your state's environmental agency website. If you use well water, you conduct your own tests through local health departments or certified labs. The diagram below from the 5th National Climate Assessment shows how climate change affects water quality:

- ⚡ Rising temperatures increase air and water temperatures.

- ⚡ Sea levels rise as ice caps melt and ocean water expands.

- ⚡ Shifts in rainfall patterns lead to more extreme hurricanes, floods, and droughts.

Water quality is degraded by:

Sediment and Debris in Waterways: Intense rainfall and flooding erode soil and transport sediments, along with pollutants in the soil, into rivers and streams.

Disinfection By-Products in Treated Water: Just as air contains unseen pollutants, water can too. When we disinfect drinking water with chemicals such as chlorine to eliminate harmful microorganisms like bacteria, viruses, and protozoans, those chemicals can react with natural substances in the water. This reaction creates toxic compounds called disinfection byproducts (DBPs), which are linked to a higher risk of bladder and colorectal cancers.

Pathogens and Toxic Algal Blooms: Warmer temperatures and nutrient-rich runoff promote harmful algal blooms that release toxins and harbor pathogens.

Groundwater Salinity: Rising sea levels can lead to saltwater infiltrating freshwater aquifers, increasing groundwater salinity.

Sewer Overflows and Pollutant Runoff into Water Sources:
Heavy precipitation overwhelms sewer systems, leading to overflows
that discharge pollutants into water bodies. Many cities use combined
sewer systems that collect both stormwater and wastewater in the same
pipes. When heavy rain overwhelms these systems, untreated sewage
spills into waterways, creating *combined sewer overflows (CSOs)*. This
pollutes drinking water sources and increases the risk of disease.

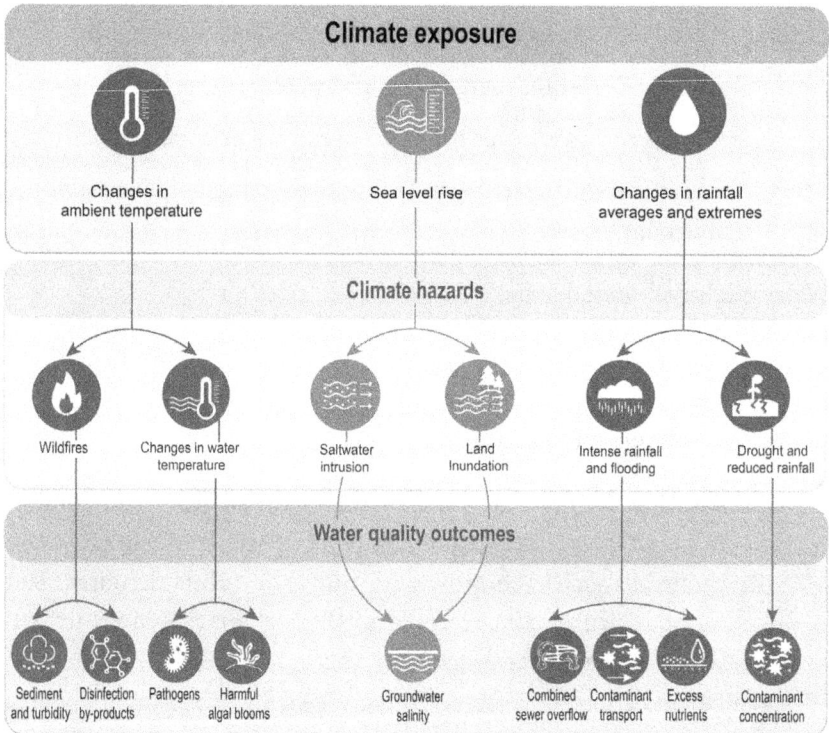

Source: U.S. Global Change Research Program, Fifth National Climate Assessment, Chapter 4:
Water—All Figures (2023), https://web.archive.org/web/20250630050556/https://nca2023.global-
change.gov/chapter/4/#fig-4-2.

Health Impacts of Floods

Drowning	The primary risk in all flood situations, especially in flash floods and fast-moving waters. Even shallow water can sweep a person off their feet.
Injuries	Floating or submerged debris, unstable buildings, and hazardous materials in the water cause cuts, punctures, and other severe injuries.
Hypothermia	Prolonged exposure to cold water or low temperatures during a flood can lead to hypothermia, especially if you're stranded without shelter.
Waterborne Diseases	Floodwaters contain pathogens like E. coli, Salmonella, and leptospirosis. Contaminated flood waters can carry sewage, livestock waste, and other hazardous materials.
Mental Health Issues	The trauma associated with floods creates stress, anxiety, depression, and PTSD.
Displacement	Flood-induced displacement interrupts medical care, worsens chronic health issues, and exposes people to new health risks in temporary shelters or unfamiliar environments. As Christiana Figueres explains, "Because multiple disasters are often happening simultaneously, it can take weeks or even months for basic food and water relief to reach areas pummeled by extreme floods."[2]
Household and Industrial Waste	Household cleaners, medical waste, and industrial chemicals can leach into floodwaters, presenting additional health risks.
Coal Ash and Heavy Metals	Coal ash and heavy metals contain harmful elements like arsenic, chromium, and mercury. These are toxic to humans and ecosystems.
Animals	Rodents, snakes, and other animals displaced by flooding can present additional threats.
Downed Power Lines	Downed power lines create electrocution risks, especially in areas where waters remain high.

Understanding Flood and Hurricane Alerts: Warnings and Watches

Source: National Weather Service, Flood Watch vs. Warning infographic, Weather-Ready Nation. https://www.weather.gov/images/rnk/outdoors/floodwarningvswatch.png.

A *Flood Watch* is issued when weather conditions suggest that flooding might occur.

⚡ Stay alert and prepare.

⚡ Monitor local weather reports and official alerts through radio, TV, websites, or mobile apps to stay informed.

⚡ Take precautions by moving valuables and electronics to higher levels in your home and securing outdoor items that could be swept away by rising waters.

⚡ Ensure your emergency kit is stocked with essentials, including non-perishable food and water to last at least three days, flashlights, extra batteries, and a first aid kit.

⚡ Familiarize yourself with the safest route to higher ground or designated shelters in case conditions worsen and evacuation becomes necessary.

A *Flood Warning* means that flooding is either imminent or actively occurring, and immediate action is necessary to ensure your safety.

⚡ If authorities instruct evacuation or you notice water levels rising rapidly, leave at once to avoid being trapped.

⚡ If evacuation is not required but flooding is expected, move to the highest floor of your home to stay above rising waters.

⚡ Never walk, drive, or swim through floodwaters. Just six inches of moving water can knock you down, and a foot of water can sweep away a vehicle.

⚡ Stay connected by using your phone or radio to receive updates from local authorities, as flood alerts provide crucial information for a timely response.

⚡ Lastly, inform loved ones of your safety and location to ensure they know you are secure.

HURRICANE WATCH

A Hurricane Watch is issued when hurricane conditions are *possible*, typically within 48 hours.

Prepare your home by boarding up windows and moving loose items indoors. Have an emergency supply kit ready.

Hurricane Winds Possible.

HURRICANE WARNING

A Hurricane Warning is issued when hurricane conditions are *expected*, typically within 36 hours.

Be ready to seek shelter in a sturdy structure or evacuate if ordered.

Hurricane Winds Expected!

weather.gov

Source: National Weather Service, Hurricane Watch vs. Warning infographic, Weather-Ready Nation, Weather.gov, https://www.weather.gov/images/wrn/Infographics/2024/240908-Hurricane-Watch-Warning.png.

A *Hurricane Watch* indicates hurricane conditions (sustained winds of 74 mph or higher) are possible within 48 hours. Monitor updates closely and prepare emergency kits and evacuation plans.

A *Hurricane Warning* signals that hurricane conditions are expected within 36 hours. Finalize your preparations and follow evacuation instructions from local authorities.

> As extreme weather becomes more frequent, there's a risk of becoming desensitized to warnings and watches. The word "unprecedented" is used so often that it no longer feels unusual. Just because extreme events are happening more frequently doesn't mean we should become accustomed to warnings. When you see a warning, remember its purpose: to keep you informed and prepared. That's the time to assess its severity and ensure you have a clear plan.

Coastal Communities

Approximately 40 percent of the U.S. population resides in coastal areas, including major urban centers such as New York and Miami. In these regions, rising sea levels and saltwater intrusion compromise freshwater supplies, pushing seawater into freshwater sources and making drinking water unsafe.

Floodwaters carry fertilizers, pesticides, and other pollutants into local waterways, while runoff from wildfires deposits ash and debris. Coastal farms face challenges as chemicals from fertilizers and pesticides pollute nearby water bodies. This poses a threat to both crops and aquatic ecosystems.

Higher water temperatures lead to the growth of harmful bacteria and toxic algae. Water treatment plants are struggling to cope with the increased debris resulting from floods and wildfires.

As sea levels continue to rise, coastal communities face an additional risk: toxic substances buried in the soil may become exposed at the surface. Twenty-five percent of the world's 460 operational commercial nuclear reactors are located along coastlines, with *nine situated in the U.S.* Many of these facilities were built just 30 to 70 feet above sea level, putting them in a vulnerable position as climate change intensifies.[3]

Before a Flood

Develop an emergency plan that includes pets. Familiarize yourself with local evacuation routes. Maintain a ready supply of essentials such as non-perishable food, water, and medical supplies.

Safeguard important documents. Store critical documents in a water-proof container and create digital copies.

Protect your home. Prepare your home by moving valuables to higher levels, clearing drains and gutters, installing check valves in plumbing to prevent sewage backups, and considering a sump pump with a battery backup for emergencies.

During a Flood

- Evacuate immediately if advised. Rising waters can trap residents, turning minor inconveniences into life-threatening situations.

- Tune in to local TV and radio stations that broadcast emergency alerts. These provide real-time updates about flood conditions, evacuation orders, and safety measures.

- A NOAA Weather Radio provides continuous updates on weather conditions, including flood warnings and alerts. Ensure you have a battery-powered or hand-crank NOAA radio in case of power outages.[4]

- ⚡ Many communities offer text or email alert systems for emergencies. Register at https://local.nixle.com/register/ or check your local town or city's website.

- ⚡ Use mobile apps like FEMA's or the American Red Cross Emergency app to receive updates and warnings.

- ⚡ For reliable updates, follow your state and local emergency management agencies and the National Weather Service on social media platforms. List of State Emergency Management Agencies: https://www.usa.gov/state-emergency-management.

Climate Cat Stays Prepared, Calm, and Equipped

Avoid Floodwaters

Floodwaters can conceal fast-moving currents and submerged hazards. Never attempt to walk, swim, or drive through floodwaters. An important saying is: *turn around, don't drown.*

On August 19, 2024, torrential rainfall caused devastating flooding in Connecticut, claiming the lives of two women stranded in separate vehicles along the Little River. Firefighters were attempting to rescue one of the victims, but the rushing water swept her away before they could reach her. The other woman managed to exit her car and cling to a sign, but the current's strength proved overwhelming.[5]

The *New York Times* article "How to Escape Your Car in a Flood" provides safety tips for avoiding life-threatening situations during vehicle submersion or flooding. It emphasizes the importance of staying calm and acting quickly, as cars can become fully submerged in minutes. Key steps include immediately unbuckling your seatbelt, unlocking doors, rolling down or breaking windows, and exiting the vehicle.[6]

After a Flood

Only return to your property when authorities say it's safe. During cleanup, wear protective clothing and use appropriate masks for mold.

Never touch electrical equipment if it is wet or you are standing in water. If it's safe, turn off the electricity in your home to prevent an electric shock.

Floodwaters are often polluted and may contain harmful pathogens. Avoid contact whenever possible, and use protective gear if necessary.

Take health precautions during cleanup. Remove wet contents and dry areas within 24 to 48 hours to prevent mold growth. Use safe drinking water and properly dispose of any contaminated cleaning materials.

Mold

Mold can quickly become a significant health risk after a flood.

Mold releases spores and mycotoxins into the air, irritating the eyes, nose, throat, and skin. It thrives in damp environments in homes, buildings, and debris. Mold growth begins within 24 to 48 hours of water exposure and spreads rapidly in porous materials like drywall, insulation, and carpeting.

Prolonged exposure can lead to a range of health issues, including respiratory problems, allergic reactions, and asthma symptoms. For individuals with weakened immune systems or preexisting conditions, exposure to mold may cause severe infections.

Mold exposure can have significant cognitive and neurological effects, particularly in cases of prolonged exposure. These symptoms are often misdiagnosed. Be alert for symptoms such as memory and concentration problems, headaches, mood changes, sleep disturbances, fatigue, dizziness, vertigo, tingling, numbness, or pain, and heightened sensitivity to smells, sounds, or light.

POWER OUTAGES

Climate change is about feeling powerless: the loss of the stable world we once knew, the loss of our homes, and the loss of electricity. Between 2013 and 2023, the ten most extensive power outages in the United States were all caused by severe weather events, including hurricanes and winter storms.

I've lived through two outages lasting more than a week in Connecticut. During one miserable winter, I had hot water (from my old gas water heater) but no heat. In 2012, Hurricane Sandy impacted 8 million people across 15 states. I experienced it.

Hurricane Irma in 2017 left 3.5 million Floridians without power. In 2024, Hurricane Helene caused power outages for millions across the Southeast. From 2000 to 2023, Texas experienced the highest number of major power outage events, with 172 incidents.

Other states with high outage counts include Michigan, California, and North Carolina. Beyond extreme weather, the other causes of power outages are aging infrastructure, demand surges, and even deliberate attacks on power grids.

Extreme weather increases electricity use, overloading the power grid. Think millions of air conditioners running on high, day and night. Demand is expected to continue increasing as temperatures rise. Data centers for AI will also consume vast amounts of electricity.

The backbone of the U.S. energy grid, including transmission lines, substations, transformers, and control systems, was built in the mid-20th century with an expected lifespan of 40 to 50 years. Today, most U.S. transmission lines are over 25 years old. Some date back to the 1950s and 1960s, built for a different time, when the nation's population was smaller, industrial needs were lower, and homes lacked today's electronics, EV chargers, or air conditioning demands.

CLIMATE CHANGE IS ABOUT FEELING POWERLESS: THE LOSS OF THE STABLE WORLD WE ONCE KNEW, THE LOSS OF OUR HOMES, AND THE LOSS OF ELECTRICITY

Health Risks of Power Outages

⚡ Outages affect home medical equipment, including CPAP machines, ventilators, oxygen systems, and dialysis machines.

⚡ Perishable foods and temperature-sensitive medications spoil without refrigeration, increasing the risk of foodborne illness and jeopardizing the health of individuals who rely on these medicines.

⚡ Water treatment and supply systems fail, leading to an increase in waterborne diseases.

⚡ Heating and cooling systems fail, causing heat-related issues, such as heat stroke and hypothermia from exposure to cold temperatures.

⚡ Emergency responders and healthcare facilities are disrupted, with pediatric clinics particularly vulnerable.

Before an Outage

Begin by developing a power backup plan for your essential needs, including lighting, refrigeration, medical equipment, and specialty medications. From personal experience, no one is prepared unless they've experienced a long-term outage. Everyone's a beginner, from knowing where their flashlight is, to battery backups, to knowing what to eat. Think this through:

⚡ How will you cook?

⚡ What are your lighting sources?

⚡ Where will you go?

⚡ What are your essential powered devices?

⚡ What are your possible sources of power?

⚡ What health risks will you and your family face during a long outage?

Power-Dependent Devices

Power-dependent devices require backup batteries or portable generators for devices like CPAP machines or oxygen concentrators. Register with utility companies for priority power restoration programs. This typically involves contacting your utility provider about your medical needs.

Visit the utility's website or call their customer service line and ask about their Medical Baseline or Critical Care programs. To maintain eligibility, you may need to fill out an application form, provide documentation from your healthcare provider, and update your registration annually. Some utilities offer free portable backup batteries.

Register for Local Planned Power Outage Alerts

During times of high wildfire risk, such as extreme heat, dry conditions, and strong winds, California's Pacific Gas and Electric Company (PG&E) may proactively shut off power to specific areas to prevent power lines from sparking wildfires.[1] These shutoffs can last from several hours to several days, depending on the conditions and the repairs needed. Oregon, Washington, Colorado, and Texas utilities also conduct wildfire-prevention shutoffs and allow customers to register for alerts.

Ensure Reliable Lighting

To ensure reliable lighting, start with basic readiness by keeping a few essential supplies, such as headlamps, flashlights, and LED lanterns. Avoid using candles, as they can cause fires if something goes wrong.

Explore Alternative Power Sources

Explore alternative power sources like generators or solar-powered chargers and test these systems to ensure they operate when needed. Portable chargers and battery packs keep phones, tablets, and laptops powered. Surge protectors and uninterruptible power supplies protect electronics from damage during power fluctuations.

Store Enough Drinking Water

Store enough drinking water for at least three days, and consider collapsible containers. If warnings are issued, fill bathtubs and sinks with water for non-drinking needs.

Assemble an Emergency Kit

The kit should include a minimum three-day supply of water and non-perishable food, medical supplies, prescription medications, a first aid kit, and spare glasses or contact lenses.

Add power-dependent items such as battery-powered fans and communication devices to enhance comfort and connectivity during an outage.

Consider the specific needs of infants, seniors, pets, and people with disabilities. Pack supplies such as baby formula, pet food, and comfort items.

Weather Radio

A NOAA Weather Radio is another essential item, providing emergency updates when cell phone networks fail.[2] It uses a 24/7 broadcast system that provides real-time weather information, forecasts, and National Weather Service (NWS) emergency alerts. It covers everything from daily weather updates and radar conditions to severe weather watches and warnings specific to your area.

The radio delivers critical emergency alerts for tornadoes, flash floods, hurricanes, wildfires, winter storms, and even non-weather-related events, such as AMBER Alerts or chemical spills. Key features include *Specific Area Message Encoding (SAME)*, which filters alerts by location, and a tone alarm that activates even when muted. It's considered one of the fastest and most reliable ways to receive warnings, often outpacing mobile apps, television, or local radio.

Because it operates independently of cell service or the internet, it remains a critical safety tool during disasters. Known as the "smoke detector for weather," a NOAA Weather Radio costs anywhere from $20 to $100, depending on the model.

During an Outage

⚡ Do not use gas stoves, ovens, or charcoal grills indoors for heating. Protect your household from carbon monoxide risks.

⚡ Install carbon monoxide detectors with battery backups and test them frequently.[3]

⚡ Follow fire safety practices, such as keeping flammable materials away from heat sources and opting for flashlights instead of candles to prevent accidents.

⚡ To avoid carbon monoxide poisoning, always operate generators outdoors, at least 20 feet from any building and away from windows.

⚡ To prevent food spoilage, keep refrigerators closed and use a thermometer to monitor temperatures. Discard any food that has been left above 40°F for more than two hours.

⚡ Research the locations and operating hours of nearby emergency cooling or heating centers and get to these facilities if needed. One example is Chicago, which operates six designated cooling centers during extreme heat. When heat advisories are issued, libraries, senior centers, park district buildings, and even CTA buses are also used as temporary cooling stations. Phoenix runs a Heat Relief Network with over 100 locations, including community centers, churches and libraries, providing water, shade, and A/C. Detroit opens multiple warming centers operated by shelters and nonprofits in coordination with the city during cold weather emergencies.

⚡ Utility providers offer information on power restoration and available community resources. Some utilities have platforms that provide outage maps and related services.

⚡ Dial 311 for non-emergency help. In many cities, 311 can connect you to information on local services, including emergency shelters.[4]

EXTREME COLD

Prepare for the unexpected.

Just as no one expected the Spanish Inquisition,[1] no one expected the cold fronts of Siberia to crash through a southern barbecue in Texas. Climate change causes the weather in one part of the world to appear unexpectedly in another.

Connecticut residents are now wading through floods, while Alaskans are sweating through unprecedented heat waves. Each region is forced to adapt to weather it never planned for. California, accustomed to managing wildfires, now faces tropical storm systems charging up the coast. Phoenix, which shrugs off 115-degree days, is paralyzed by flash floods. The East Coast, where winter emergencies are as routine as morning coffee, now grapples with weeks of triple-digit heat waves that turn subway stations into saunas.

We looked at the impact of rising temperatures and heat waves, but extreme weather also brings frequent and intense cold. What does global warming have to do with freezing temperatures?

One driver is the weakening of the *polar vortex*. This is defined as a large area of low-pressure, cold air that typically hovers over the Arctic. As the Arctic warms faster than other parts of the planet, the temperature difference between the Arctic and mid-latitudes shrinks. This reduction in temperature difference causes the polar vortex to become unstable and shift southward.

Sudden and extreme cold outbreaks then occur in areas that have never experienced such conditions before. For example, in 2014, the polar vortex brought near record-low temperatures to parts of the southern United States, with temperatures plunging in Georgia and Florida.

Extreme weather has disrupted the jet stream, a fast-moving air current that plays a major role in shaping weather patterns. When the jet stream slows down and becomes "wavier," it can result in prolonged cold spells, as frigid air remains stationary over specific regions. These cold spells create health risks, particularly in states that never experience intense cold. In

February 2021, a severe winter storm swept across Texas, leaving millions without power and resulting in a surge in hypothermia cases.

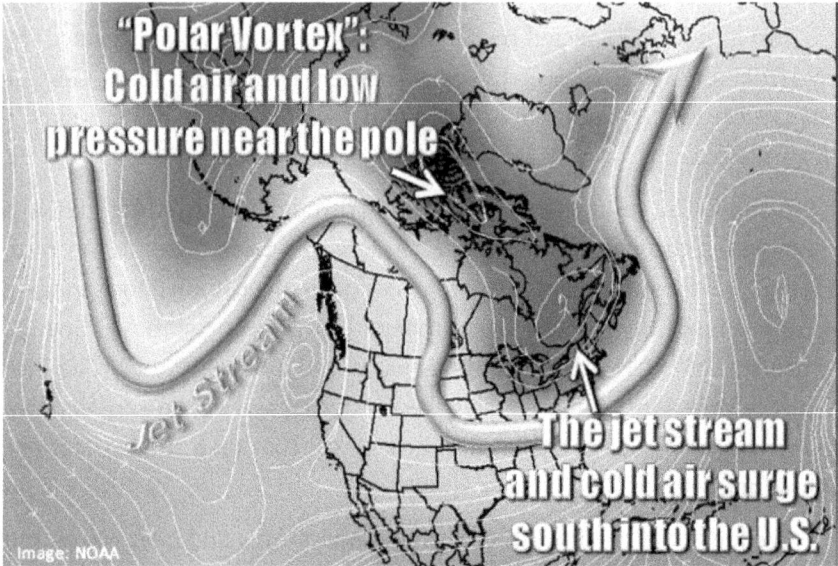

Source: National Weather Service, "What is the Polar Vortex?" infographic, https://www.weather.gov/safety/cold-polar-vortex.

Health Risks of Extreme Cold

Hypothermia happens when your body loses heat faster than it can generate it, causing your core temperature to drop to a dangerously low level. Symptoms include persistent shivering, confusion, impaired memory, slurred speech, exhaustion, and, in severe cases, loss of consciousness. If left untreated, hypothermia can cause organ failure and death. According to the CDC, in the U.S., hypothermia contributes to 1,300 deaths annually.

Frostbite freezes the skin and underlying tissues. It affects the fingers, toes, nose, ears, cheeks, and chin. These areas become numb, hard, and discolored in severe situations, ranging from pale to bluish-black. In extreme cases, frostbite leads to permanent tissue damage and amputation. During the 2021 Texas freeze, hospitals reported numerous cases of frostbite, some of which required amputations due to delayed treatment.

EACH REGION
IS FORCED
TO ADAPT
TO WEATHER
IT NEVER
PLANNED FOR.

Cardiopulmonary strain occurs in cold conditions, which require the heart to work harder to maintain body temperature. This increases the risk of heart attacks, particularly in individuals with preexisting heart conditions. Cold air irritates the respiratory system, causing airway constriction and breathing difficulties. This strain on the lungs worsens asthma and chronic obstructive pulmonary disease (COPD). Breathing in cold, dry air can trigger an asthma attack. Research indicates that cold weather is associated with a 34 percent increase in heart attack rates.[2]

Dehydration can occur because cold air is dry, and breathing it in increases water loss through respiration. Cold weather can reduce a person's thirst response by up to 40 percent. Blood vessels constrict to reduce heat loss, redirecting blood flow to the body's core to protect vital organs. This tricks the body into feeling hydrated, which results in drinking less water and an increased risk of dehydration.[3] Severe dehydration causes dizziness, rapid heart rate, and impaired kidney function. Many people underestimate their fluid needs during cold weather, compounding this risk.

Weakened immune response can result from prolonged exposure to cold, which then impairs the body's defenses and increases susceptibility to colds, flu, and other respiratory infections. Flu cases often spike following severe cold snaps.

Vulnerable populations are especially at risk, as we've seen throughout this book. Infants and young children lose heat more quickly due to their smaller body size. Seniors are more likely to have medical conditions that heighten their sensitivity to cold, and many experience social or physical isolation, which can make it difficult to access help.

People with chronic health conditions, such as heart and lung diseases, diabetes, or impaired mobility, are vulnerable to cold-related complications. Certain medications, including sedatives, beta blockers, and diuretics, can further impair the body's ability to regulate temperature.

People experiencing homelessness are at high risk, as the lack of proper shelter and warm clothing significantly increases the likelihood of cold-related illnesses and fatalities. As public health researcher Kathryn Lane and her colleagues warn in their 2018 study of cold-related illness and death in New York City, "Unsheltered homeless individuals, people who use substances and become incapacitated outdoors, and older adults with medical and psychiatric conditions without home heat are most at risk."[4]

Spending time outdoors in cold environments is dangerous even for healthy individuals, as extended exposure and inadequate preparation can lead to life-threatening conditions.

Areas That Are Unprepared

States that rarely experience severe cold weather are often completely unprepared, leading to major public health issues. For instance, during the 2014 "polar vortex," near record-low temperatures were reported as far south as Georgia and Florida, resulting in widespread disruptions and numerous cold-related injuries. Hospitals in these regions faced a surge in patients with hypothermia, frostbite, and respiratory issues.

In some cases, residents resorted to unsafe heating practices, such as using indoor grills and gas stoves, which led to increased carbon monoxide production. More than 1,400 cases of carbon monoxide poisoning were reported during the Texas freeze.

Common Signs of Carbon Monoxide Poisoning

- Headache (one of the earliest symptoms of carbon monoxide (CO) exposure)
- Dizziness, weakness, and fatigue
- Nausea and vomiting
- Shortness of breath
- Difficulty breathing even with minimal activity
- Confusion or mental impairment
- Blurred vision or loss of consciousness

If you suspect carbon monoxide poisoning, it's essential to move to fresh air immediately and call emergency services.

Additionally, icy roads and unprepared transportation systems delayed emergency responses, which worsened health outcomes for those in need of urgent medical attention.

*"Watch the Weather, Warm the House,
and Guard Your Health"*

CLIMATE CAT'S COLD WEATHER TIPS

Stay Informed About the Weather: Just as you monitor weather forecasts for heat, air quality, and hurricanes, pay attention to alerts regarding cold. Early awareness can save lives and reduce the severity of cold-related incidents.

Dress Appropriately for the Cold: Particularly important for states that are usually *hot*. Wear layers of loose, warm clothing. Select outer layers that are both windproof and water-resistant to retain heat and prevent moisture. Avoid cotton because it retains moisture and can increase heat loss. Pay close attention to extremities by wearing insulated gloves, hats, and scarves.

Recognize the Warning Signs of Cold-Related Illnesses:
Seek warmth immediately if you notice severe shivering, numb
extremities (fingers, toes, nose, ears, cheeks, and chin), disorientation, or extreme fatigue, which may indicate frostbite or
hypothermia. Early intervention is crucial in preventing severe
outcomes. For example, during a cold snap in Chicago in 2019,
rapid medical intervention saved multiple lives after frostbite was
detected early in homeless populations.

Hypothermia: Move the person to a warmer place. Remove wet
clothes, wrap them in a blanket, cover their head to warm them
slowly.

Check on Vulnerable People: Check on neighbors to ensure
their safety and access to warmth. Community support can prevent
isolated incidents from becoming tragedies.

Hydrate: Maintaining hydration is essential in cold weather.
Carry water during outdoor activities and monitor for signs of
dehydration, such as dark urine or fatigue.

Limit Physical Exertion: Avoid overexertion. Strenuous activity
increases the risk of heart attacks. Snow shoveling, for example,
accounts for over 100 cardiac-related deaths annually in the U.S.
during winter.

Use Heating Safely: Don't use gas stoves, ovens, or grills for
home heating. They pose significant fire and carbon monoxide
risks. Instead, use space heaters with automatic shut-off features
and never leave them unattended.

Identify Community Resources: Know where local warming
centers and emergency shelters are. In 2021, Chicago's centers saved
lives by offering refuge during extreme cold.

HOW TO BE PREPARED

CHAPTER 4

Plan and Protect

"Those who plan in peace move swiftly in crisis."
—Climate Cat

WEATHER LITERACY AND ALERTS

Weather literacy has become an essential skill in the twenty-first century. We live in a time of unpredictable weather patterns, information overload, and crisis after crisis. Our lives, livelihoods, and safety depend on understanding the progression of wildfires, hurricanes, and heat waves.

Ignoring or politicizing extreme weather will not make it disappear. Reliable weather data is a key to our survival.

This chapter explores the tools used to understand weather, from apps and maps that provide real-time updates to advanced models that predict long-term impacts. We'll review how to interpret weather reports, track hurricanes and heat, and set up personalized alert systems to protect you and your family.

Weather vs. Climate

Let's look at the difference between weather and climate:

⚡ Weather refers to the day-to-day conditions around us, including rain, snow, heat, and hurricanes.

⚡ Climate describes weather patterns over extended time frames (years or decades) and helps us understand trends like hotter summers, rising sea levels, and more intense storms.

Understanding the weather allows us to react to immediate threats. Understanding climate helps us prepare for longer-term impacts: frequent floods, the spread of more diseases by insects, and adapting to a hotter world.

Interpreting Weather Reports

Weather reports help you plan your day and stay safe during extreme conditions:

Temperature Forecasts: Look for daily highs and lows, and pay attention to the heat index, which takes humidity into account to provide a "feels like" temperature.

Precipitation: Check the likelihood (shown as a percentage), timing, and expected accumulation of rain or snow.

Wind Conditions: Monitor wind speed and direction, which is especially important during fire season or severe storms.

Severe Weather Alerts: Pay close attention to watches (indicating a weather event is likely to occur) and warnings (notifying you of severe weather that is imminent or already occurring).

The Air Quality Index: Indicates the levels of pollutants in the air; it is essential to know if you have any respiratory conditions.

Reading Your Weather App

Hourly and Daily Forecasts: Shows temperature, precipitation, and wind predictions over time.

Radar Imagery: Visualizes precipitation patterns and storm movements in real-time.

Barometric Pressure: Falling pressure often signals the approach of storms; rising pressure typically indicates clearing conditions.

Humidity Levels: High humidity can increase heat stress and alter the perceived temperature.

Dew Point: Indicates the temperature at which air becomes fully saturated with moisture, leading to condensation. High dew points (above 65°F) mean the air feels muggy and can be oppressive, making it harder for sweat to evaporate and cool the body.

Visibility: Indicates how far you can see clearly, which is vital during foggy or heavy rain conditions.

Alerts Banner: Usually highlighted in red or orange for severe weather warnings.

UV Index: Indicates the strength of ultraviolet radiation, which will tell you if you need sunscreen, hats, sunglasses, or shade.

UV Index Risk Levels and Protection Guidelines

UV INDEX	RISK LEVEL	PROTECTION ADVICE
0-2	Low	Safe for most. Minimal protection is needed.
3-5	Moderate	Use sunscreen, a hat, and sunglasses if you are outside.
6-7	High	Apply SPF 30+. Wear protective clothing and sunglasses.
8-10	Very High	Extra protection. Limit midday sun. SPF 30+, reapply often.
11+	Extreme	Avoid sun exposure. Full protection is critical.

ESSENTIAL WEATHER AND CLIMATE RESOURCES

Below are charts of helpful websites and apps for becoming weather literate, organized by climate impact. These will help you monitor conditions, receive alerts, and understand both short-term weather and long-term climate trends. *Additional resources can be found in the Appendix.*

I've included many links intentionally.

In an era when trustworthy information is becoming increasingly difficult to find and even basic science is attacked, I'd rather overshare than leave you searching. Don't feel like you have to explore everything at once. Skim the list, bookmark what's useful, and return when needed.

General Weather Literacy

- **National Weather Service (NWS)** • https://www.weather.gov/

  The official U.S. source for weather forecasts, alerts, and specialized predictions for aviation, marine, and fire weather. Many commercial weather platforms source their data from the National Weather Service, then enhance it with their own radar systems, local weather stations, and predictive models.

- **The Weather Channel** • https://weather.com/

  Provides hourly, daily, and weekly forecasts, interactive radar, alerts, and weather maps for tracking severe weather conditions. Integrates AQI data with weather forecasts.

- **AccuWeather** • https://www.accuweather.com/

  Known for its precise localized forecasts, utilizing the "RealFeel" index that factors in environmental conditions such as humidity and wind. Offers a "MinuteCast" feature, real-time radar, and severe weather alerts.

- **Weather Underground** • https://www.wunderground.com/

  Uses a network of local weather stations to deliver hyper-local forecasts and interactive maps, offering real-time data for specific neighborhoods. Excellent for monitoring microclimate variations.

- Windy • https://www.windy.com/

  Offers sophisticated visualization of weather patterns including temperature, humidity, and air quality. Provides a wet bulb globe temperature layer for tracking heat stress with a detailed and user-friendly interface.

- Tomorrow.io • https://www.tomorrow.io/consumer-app/

  Specializes in "micro-weather" forecasting, offering hyper-local weather updates and predictive weather conditions tailored to your location.

- FEMA National Risk Index • https://hazards.fema.gov/nri/

  This interactive tool provides a comprehensive assessment of natural hazard risks across the U.S., evaluating and mapping hurricanes, earthquakes, floods, and wildfires.

Air Quality Monitoring

Poor air quality can trigger respiratory problems, especially if you have asthma, COPD, or other lung conditions. Key factors to watch include particulate matter, ozone, and air quality index (AQI) ratings. Ozone levels tend to rise on hot, sunny days. It's important to check smoke forecasts during wildfire season, as smoke worsens air quality over very large areas.

- AirNow (EPA) • https://www.airnow.gov/ • https://www.airnow.gov/airnow-mobile-app/

  Real-time data on particle pollution, ozone levels, fire weather alerts, and smoke maps. Offers a Fire and Smoke Map to monitor wildfire activity and smoke conditions.

- AirVisual/IQAir • https://www.iqair.com/us/air-quality-monitors/air-quality-app

  Global air quality updates with detailed health recommendations and breakdowns of specific pollutants. Features include pollution forecasting, historical data analysis, and customizable alerts when AQI reaches unhealthy levels.

- PurpleAir • https://map.purpleair.com/air-quality-standards-us-epa-aqi

  A crowd-sourced network of air quality sensors providing hyper-local air quality data worldwide. It is particularly useful for tracking wildfire smoke and particulate matter in real time.

- Pollen.com • https://www.pollen.com/

  Helps allergy sufferers track pollen forecasts by location with information on specific allergens like tree pollen and ragweed.

Heat Monitoring and Alerts

Heat is one of the deadliest forms of extreme weather. Sometimes, we can prepare for a heat wave days in advance; other times, it strikes suddenly. Key factors to monitor include temperature forecasts and trends, the heat index, wet bulb temperature, and the duration of heat waves.

- **Zelus WBGT** • https://zelusports.com/
  Designed to monitor wet bulb globe temperature (WBGT) using localized weather data, including humidity and solar radiation. Used by schools and universities to monitor conditions for athletics, fieldwork, and other outdoor programs.

- **CDC Heat Safety Tool App** • https://www.cdc.gov/niosh/heat-stress/communication-resources/app.html
  The OSHA-NIOSH Heat Safety tool provides hourly heat index forecasts and guidance on preventing heat-related illnesses.

- **Heat.gov** • https://www.heat.gov/
  An initiative from the National Integrated Heat Health Information System (NIHHIS), with real-time heat risk data and health, interactive maps, and community planning guidance.

- **NOAA HeatRisk Tool** • https://www.wpc.ncep.noaa.gov/heatrisk/
  Provides a color-coded risk index to forecast heat-related risks up to seven days in advance.

- **Tempest Heat Alert** • https://heatalert.app/
  Focuses on health impacts, with specific recommendations for different activity levels and vulnerability factors. Includes wet bulb temperature monitoring.

Fire Monitoring and Safety

Fire forecasts highlight the combination of heat, dry conditions, and strong winds that increase the likelihood of wildfires. Pay close attention to Red Flag Warnings, which signal periods of heightened fire danger. Tracking smoke plumes can help you understand the impacts of air quality and potential threats to your area. Stay alert for evacuation notices to ensure you have enough time to respond quickly if a wildfire approaches.

- **EPA AirNow Fire Preparation, Response and Safety** • https://www.airnow.gov/wildfires/

  Tips for staying safe before, during, and after a fire.

- **AirNow Fire and Smoke Map** • https://fire.airnow.gov/

  Real-time air quality and smoke updates from the EPA and U.S. Forest Service.

- **Frontline Wildfire App** • https://www.frontlinewildfire.com/frontline-wildfire-app/

  Offers wildfire alerts, evacuation info, and emergency planning tips.

- **Ready.gov Wildfire Resources** • https://www.ready.gov/wildfires

  Helps you understand wildfire risks and plan evacuation routes.

- **InciWeb Incident System** • inciweb.wildfire.gov/

  Provides real-time updates on active wildfires, interactive maps, and evacuation notices.

Flooding

Tools for monitoring and assessing flood risk include precipitation forecasts and flood stage measurements for rivers and streams. Flash flood warnings are essential for areas vulnerable to sudden, intense flooding, while storm surge predictions are critical for coastal regions. *You can find additional flood resources in the Appendix.*

- **NOAA Coastal Flood Exposure Mapper** • https://coast.noaa.gov/floodexposure/

  An interactive map that shows areas at risk of coastal flooding from sea-level rise, storm surge, and inundation.

- **FEMA Flood Map Service Center** • https://msc.fema.gov/portal/home

  Lets you search floodplain maps by address, helping you understand flood risks and insurance requirements.

- **Climate Central's Surging Seas** • **Risk Finder** https://riskfinder.climatecentral.org/

  Projects sea-level rise impacts and identifies high-risk zones.

Hurricane and Storms

When hurricane clouds start swirling, these apps put the storm's track, intensity, and life-saving alerts on your phone.

- MyRadar Weather • https://myradar.com/

  Real-time tracking of storms and hurricanes featuring animated radar maps and storm-tracking capabilities.

- Red Cross Emergency App • https://www.redcross.org/get-help/how-to-prepare-for-emergencies/mobile-apps.html

  Designed for disaster preparedness with real-time alerts for natural disasters, including heat, hurricanes, wind, tornadoes, and wildfires.

Climate Projections and Long-Term Planning

To help you prepare for the longer-term impacts of climate change, several interactive tools offer maps, graphs, and projections that visualize future scenarios.

- Climate Explorer • https://storymaps.arcgis.com/stories/15421d03a5744cf2bf83c9f016eb458c

  Interactive maps and graphs show climate projections by U.S. county and decade, illustrating possible futures based on different emission scenarios.

- NOAA Climate Prediction Center • https://www.cpc.ncep.noaa.gov/

  Authoritative seasonal forecasts, drought assessments, and long-term climate outlooks with interactive tools.

Global Weather Monitoring

For a worldwide view, whether you're checking tomorrow's forecast in Tokyo or tracking a cyclone off Bermuda, these tools keep you instantly informed about weather and disasters unfolding across the globe.

- BBC Weather • https://www.bbc.com/weather

  Provides global forecasts with detailed temperature trends, weather alerts, and daily summaries.

- Pacific Disaster Center • https://disasteralert.pdc.org/disasteralert/

  Maps that show extreme weather events occurring globally in real time.

Setting Up Your System for Alerts

Step 1: Build Redundancy

You can use the National Weather Service website for official alerts. Add a commercial weather app, such as AccuWeather or Weather.com, for general weather updates, and consider using an air quality app like AirNow or PurpleAir to monitor pollution levels. Choose a specialized app for your region's risks, such as hurricanes, flooding, or wildfires, to stay informed on local threats.

Here's an example of what my phone looks like:

Step 2: Enable Push Notifications

This is crucial for receiving real-time alerts, particularly during emergencies when every minute matters. Notifications will ensure you stay updated instantly and can take action when conditions change or hazards arise.

Step 3: Set Up Multiple Locations

Configure each app to follow all the places that matter, not just your current GPS pin. Add your home, workplace, children's schools, relatives' addresses, and any evacuation destinations or travel routes you will use. This multi-location setup lets you see where threats are building, so you can act quickly to protect yourself and the people who rely on you, no matter where they are.

Step 4: Familiarize Yourself with App Features

Before an emergency arises, take the time to explore each app's features. Learn how to access radar and satellite imagery, adjust alert options, and understand forecast details.

Emergency Alert Systems

These tools provide real-time updates, safety information, and access to critical resources during disasters or severe weather events.

- **NOAA Weather Radio • https://www.weather.gov/nwr/**

  Nationwide network broadcasting continuous weather information from NWS offices (seven frequencies within 162.400–162.550 MHz). You can find many NOAA radios on Amazon for under $30.

- **Nixle Alerts • https://local.nixle.com/register/**

  Local emergency alerts, advisories, and community information customized to your area.

- **FEMA App • https://www.fema.gov/about/news-multimedia/mobile-products**

  Real-time disaster alerts and preparedness tips based on your location, with space to store emergency contacts.

- **Smart911 • https://www.smart911.com/**

  Customizable safety profiles for individuals with disabilities that can be shared with emergency first responders.

- **State Emergency Management Agencies • https://www.usa.gov/state-emergency-management**

  Allows you to follow your local government and state emergency management agencies on social media.

- **Emergency Shelter Locator • Search for open shelters by texting SHELTER and your ZIP code to 43362 (4FEMA).**

  Text-based service to find nearby shelters during emergencies.

Weather literacy is more than knowing if it will rain tomorrow. It's understanding the complex systems that affect our daily lives and long-term safety. *Remember:* Climate information is power. Stay informed, stay prepared, and stay safe.

EMERGENCY KITS

Emergencies are unpredictable and often come with little warning. Some disasters can be modeled, predicted, or projected, while others cannot. For years, the word *prepper* conjured images of people stockpiling canned goods in underground bunkers, awaiting the apocalypse. With increasingly frequent and severe climate disasters . . . the paranoid seem prescient.

We can learn from their commitment to self-reliance, readiness, and wise preparation. So, let's embrace the best elements of the preparedness ethos, from building go bags to learning first aid to community emergency response.

In preparing for and responding to climate disasters, *bug-in* and *bug-out* refer to two different strategies you can use, depending on the severity of the disaster and your situation.

Bug-in involves staying in place and sheltering within your home or another secure location when a disaster strikes. The goal of bugging in is to create a safe and secure environment where you can weather the storm or hazard without needing to leave. Often preferred for heat waves or brief smoke incursions; less so when fire or water directly threatens your home. This is ideal in situations where remaining indoors is safer than evacuating.

When bugging-in, focus on fortifying your home and stockpiling sufficient supplies, including food, water, medical kits, and alternative power sources—so you can be self-sufficient until the threat has passed. Bugging-in is preferred if evacuation routes are unsafe or if the disaster's severity doesn't require immediate relocation.

During a blizzard or a heat wave, it may be more dangerous to venture outside or attempt evacuation than to remain indoors. Your home must be equipped to withstand the elements, and you should be prepared for an extended period of isolation.

Bug-out refers to evacuating your home or current location when staying in place becomes too dangerous. This may happen during events like flooding, hurricanes, or wildfires that render your home uninhabitable.

When bugging-out, you need a well-packed emergency kit (a "go bag") with essential items such as water, food, medication, first aid supplies, and personal documents. The goal is to be able to leave quickly and safely, typically heading to a predetermined safe location like a friend or relative's house, an emergency shelter, or an evacuation site.

Bugging-out may be required when authorities issue mandatory evacuation orders due to imminent dangers like rising floodwaters, wildfires, or toxic spills, where remaining at home could be life-threatening. A bug-out plan should include a clear evacuation route and communication strategies with loved ones, ensuring everyone knows where to go and how to stay in contact.

You can incorporate both bug-in and bug-out strategies into your climate preparedness plan. The decision between the two depends on the type and severity of the disaster, your location, and the availability of resources. Effective preparedness involves having plans and supplies for both scenarios, allowing for flexibility depending on how the situation unfolds:

To Bug-Out or Bug-In

TYPE OF DISASTER	BUG-OUT OR BUG-IN	DETAILS
Hurricanes	Bug-Out	Evacuate if you're in the storm's direct path or if you live in a flood-prone area; follow evacuation orders.
Tornadoes	Bug-In	Seek immediate shelter in a basement or interior room; evacuating is often impossible.
Floods	Bug-Out	Evacuate if you live in a low-lying or flood-prone area or are instructed to leave; move to higher ground promptly.
Wildfires	Bug-Out	Wildfires can move extremely rapidly, and waiting for official evacuation orders can sometimes be too late. Stay informed about fire progression and air quality.
Earthquakes	Bug-In	Drop, cover, and hold on during shaking; evacuate only if the building is unsafe after the event.
Severe Winter Storms	Bug-In	Stay indoors to avoid hazardous conditions; ensure adequate supplies.

Shelter-in-place (bug-in) preparations include:

✓ Designating a secure location in your home, such as a basement for tornadoes or a sealed room for air quality.

✓ Maintaining a two-week supply of essentials, including non-perishable food, water, hygiene products, and medical supplies.

✓ Investing in reliable communication tools like battery-powered radios or satellite phones to stay connected.

✓ Planning for extended periods of sheltering by incorporating entertainment and stress-relief options.

The Appendix provides a comprehensive list of items for your go bag and resources for sheltering in place (bugging-in).

Bug-Out Basics

Consider the essentials: food, water, clothing, comfort, and communication.

NOAA Radio	NOAA Weather Radio for critical emergency alerts. This is important if you can't access cell service or the internet.
Cell Phones with Chargers	Include a backup battery for emergencies.
Comfort Items	Sleeping bags, warm blankets, a change of clothes, and sturdy shoes.
First Aid Kit and Medical Supplies	A comprehensive kit with basic medical supplies: pain relievers, anti-diarrhea, antacids, and laxatives; soap, hand sanitizer, and disinfecting wipes; feminine supplies and other personal hygiene items; moist towelettes, hand sanitizer, garbage bags, and plastic ties.
Flashlight	Functional flashlight with extra batteries.
Food	Non-perishable food for several days. Canned goods, snacks, dried meat and fruits, and other shelf-stable items you enjoy.
Medications	A 30-day supply of prescription medications and other essentials. Some states allow emergency prescriptions for up to 30 days without prior approval.
Water	Water supply for hydration. One gallon per person daily for drinking and sanitation. FEMA recommends a minimum three-day water supply.

Medical Emergency Kit

One of the most important things you can do when preparing for emergencies is build a medical emergency kit tailored to your family's unique needs. Think of it as peace of mind packed in a bag, ready to help you manage unexpected situations calmly and confidently. Your kit should include essentials like medical records, medications, and first aid supplies, as well as personal touches like comfort items or foods for special dietary needs.

Every family is different, so add a small cooler for insulin, baby supplies, or a backup charger for medical devices. The key is to think ahead about what keeps your loved ones healthy, safe, and comfortable, and make sure those things are ready to go when you need them.

Medical Records:

✓ Store medical records electronically on secure cloud platforms or keep physical copies in waterproof, labeled folders. You can also save records on encrypted USB drives and store them in a secure, accessible location. Many insurers provide access to medical records through patient portals like MyChart.

Medications:

✓ Maintain an extended supply of critical medicines. Include backup inhalers, insulin, and EpiPens if needed.

✓ Create a complete list of prescribed medications, including dosages and schedules. This information can be found in your electronic health record, through your doctor's patient portal, your health insurer, or online at your pharmacy. Create a list in advance for seniors so they can carry it with them on their phones or a printed sheet. Consider adding notes on the purpose of each medication to assist others.

✓ Keep printed or digital copies of prescriptions. Store these in waterproof bags or secure digital backups to prevent loss.

✓ For medications that require refrigeration (like insulin), you may need a small cooler with ice packs. For longer emergencies, consider portable refrigeration options.

Medical Devices:

✓ Bring essential devices, such as glucose meters, blood pressure monitors, nebulizers, and CPAP machines, along with their corresponding batteries and chargers.

Basic First Aid Kit:

✓ Include bandages, antiseptics, gloves, pain relievers, and medical tape. You can expand the kit to include scissors, tweezers, and a first aid manual for quick guidance.

Sanitary Supplies:

✓ Pack hand sanitizer, soap, wipes, adult diapers, feminine hygiene products, disposable washcloths, and towels.

Masks:

✓ Include surgical or N95 masks for respiratory protection. Consider child-sized masks for families with young children and additional filters for extended use.

Baby Supplies:

✓ Include baby formula, food, and bottles. Don't forget a small bottle brush and a collapsible drying rack for cleaning feeding supplies.

Stress-Relief Tools:

✓ Pack books, puzzles, tablets, or comfort items to reduce stress. Comfort Items: Bring a lightweight blanket or pillow, along with familiar items for children, such as a favorite toy or stuffed animal.

ONE OF THE
MOST IMPORTANT
THINGS YOU
CAN DO WHEN
PREPARING FOR
EMERGENCIES IS
BUILD A MEDICAL
EMERGENCY KIT
TAILORED TO
YOUR FAMILY'S
UNIQUE NEEDS.

Communication:

✓ Include a notebook and pen for writing messages or keeping track of important information.

Emergency Contacts:

✓ Keep a written list of emergency contacts, including healthcare providers, family, and local emergency services.

The New York Times has a curated guide that recommends essential items to include in your first aid kit, ensuring you're prepared for emergencies. (https://www.nytimes.com/wirecutter/reviews/lifesaving-items-for-first-aid-kit/)

Maintaining Your Kit

Once your emergency kit is assembled, you'll need regular maintenance. Replace expired items, including food, medications, and batteries, and adjust supplies as your family's needs change.

Ensure all family members know where the kit is stored and how to access it quickly. Keep the kit in a designated location and make it readily available for quick evacuation.

You can prepare a smaller kit for work that has food, water, and other necessities for at least 24 hours, along with walking shoes. You can keep a kit of emergency supplies in your vehicle in case you become stranded.

FAMILY COMMUNICATION AND EVACUATION PLANNING

Creating a reliable emergency plan for your family is like writing a playbook for a sports team. Everyone needs to understand their role and know what to do in various situations. It's not enough to have a plan in your head; it needs to be written down, practiced, and regularly updated to account for changes in family circumstances or new risks.

This template from Ready.gov can be used to create a family emergency communication plan. It includes sections for gathering important contact information, identifying emergency meeting locations, and documenting medical information. (https://www.ready.gov/sites/default/files/2020-03/family-emergency-communication-planning-document.pdf) The Ready.gov "Plan" page guides you through creating a customized emergency plan for your household. (https://www.ready.gov/plan)

Establish Communication Methods

Choose multiple ways to stay in touch during emergencies. Text messages are often more reliable than calls during disasters. Other good backup options include social media, email, and two-way radios. Write down important phone numbers and email addresses in case phones stop working. Emergency communication tools such as satellite phones or personal locator beacons are also available for remote or high-risk areas.

Amateur radio (sometimes called "ham radio") can be used for emergency communications. When traditional communication infrastructures such as cellular networks, landlines, and wireless networks are compromised, amateur radio can provide reliable communication channels. Families can use a ham radio to maintain contact with emergency services, receive critical information, and coordinate with others. (To transmit on amateur bands in the U.S., you must hold at least a Technician-class license.) The American Radio Relay League (ARRL), established in 1914, serves as the national association for amateur radio in the U.S., offering resources and training: https://www.arrl.org/.

Designate a Central Contact

Choose someone out-of-area as your family's central point of contact. This person can relay messages between family members if local networks are overwhelmed. Family members should check in with this contact at designated intervals. Establish a secondary contact in case the primary contact becomes unreachable. Ensure that both contacts are well informed about your family's emergency plan and can assist you.

Designate safe and familiar places where your family can regroup if separated during a disaster.

✓ Establish a neighborhood and an out-of-town meeting location for flexibility, depending on the situation.

✓ Consider mapping out multiple routes to these locations and marking alternate meeting spots.

✓ If you have young children, practice walking or driving these routes together to build familiarity and confidence in reaching the destinations safely.

✓ Plan for times when family members might be separated, including clear instructions for school-aged children on how to seek help.

When you've mapped out how to communicate, the next step is building the support infrastructure: lining up helpers, tailoring for special needs, and knowing where shelters are. No one in your family should face an emergency alone or unprepared.

Establish a Network of Helpers: This should include neighbors, community groups, or friends who can provide assistance in emergencies.

Plan for Specific Needs: Tailor your communication plan to accommodate your family's needs, including medical conditions, disabilities, pets, and children. Ensure your plan accounts for necessary medications or specialized care.

CREATING A RELIABLE
EMERGENCY PLAN
FOR YOUR FAMILY
IS LIKE WRITING
A PLAYBOOK FOR
A SPORTS TEAM.
EVERYONE NEEDS TO
UNDERSTAND THEIR
ROLE AND KNOW
WHAT TO DO IN
VARIOUS SITUATIONS

Identify Local Shelters: Check the Disaster Assistance website to familiarize yourself with local emergency shelter locations. (https://www.disasterassistance.gov/information/immediate-needs/emergency-shelter)

- Shelters can be found through town websites, local churches, community centers, or organizations like the American Red Cross. Emergency shelters provide food, water, and sleeping accommodations.

- They offer facilities like charging stations for electronic devices, helping to maintain communication during a crisis. Shelters connect individuals and families with essential local services, including medical assistance, counseling, and information about ongoing recovery efforts.

Create a simple chart outlining who is responsible for what during an emergency. Write down emergency contacts, evacuation routes, and meeting points. Make sure everyone has a digital and paper copy available. Use a cloud-based storage system to keep the plan accessible even if physical copies are lost. Regularly review and update the document to reflect any changes in contact information, family circumstances, or risks.

Hold family meetings to review and practice the plan. Test communication methods, including sending texts and calling your out-of-town contact. Familiarity with the plan will help everyone stay calmer during an emergency. Schedule drills that simulate different scenarios, such as power outages, evacuations, or medical emergencies, to build preparedness and identify areas that can be improved.

Communication Tools and Technology

Establish communication plans with family members, caregivers, and healthcare providers.

Communication Tree: Designate a primary contact to relay updates to a small group, who then pass on the message. Create multiple branches for large families or communities to ensure messages spread quickly.

Internet-Based Platforms: Use WhatsApp, Signal, or Telegram when cellular networks fail, but Wi-Fi is operational. Maintain backup power sources for devices. Switch devices to low-power mode when fully charged to conserve battery life during prolonged outages.

Landlines: Copper-wire landlines may function during power outages. You can keep a corded phone as a backup.

Social Media Alerts: Platforms like Facebook and Instagram provide emergency updates and can share automated alerts from emergency feeds. Encourage family members to follow local emergency management accounts for real-time information.

Evacuation Planning and Preparation

Throughout this book, you've learned how to understand climate risks specific to your region, including fires, floods, and hurricanes. You explored how to effectively use weather apps to track extreme weather developments and set up reliable alert systems. Understanding *how* to evacuate before you need to evacuate means taking a thoughtful, step-by-step approach.

Evacuation Essentials

✓ **Is your car filled with gas?**

Keep your gas tank at least half full to avoid delays in case of an emergency.

✓ **Do you know what to do if your doctor is unavailable?**

Have a backup plan for healthcare needs. Familiarize yourself with alternatives in case primary care is unavailable.

✓ **Do you know the hospitals and urgent care centers within a 30-mile radius?**

During disasters, hospitals may be inaccessible or strained due to resource limitations. You can confirm hospital statuses through state or local health departments. Be ready to use alternatives like urgent care, community clinics, retail clinics, and telehealth services.

✓ **Do your family members have specific health needs?**

If so, pack medications, medical equipment, and special foods.

✓ **Do you know where you're going?**

Plan for where you'll stay: family, friends, shelters, or hotels.

✓ **Have you identified multiple escape routes?**

If one route becomes blocked or unsafe, know at least two different routes out of your area. Practice these routes regularly.

✓ **Did you prepare go bags for everyone?**

Prepare an individual "go bag" for each family member, containing essentials like clothing, medications, important documents, hygiene items, snacks, water, and comfort items.

✓ **Did you sign up for special needs registries?**

Many towns and cities offer registries for those needing extra assistance during evacuations. Sign up if applicable.

The Five P's of Evacuation

1. **People/Pets:** Make sure that all family members and pets are accounted for. For pets, pack carriers, leashes, IDs, and enough food and water.

2. **Prescriptions:** Gather medications, medical supplies, and clear dosage information.

3. **Papers/Plastic:** Protect essential documents (IDs, insurance info, credit and debit cards, medical records, financial details) in waterproof containers.

4. **Personal Items:** Pack clothing, hygiene products, comfort items, and electronics like phones, tablets, and chargers.

5. **Priceless Items:** Carefully select irreplaceable items such as family photos or heirlooms, if safe and feasible.

Top 10 Evacuation Mistakes to Avoid

1. **Ignoring Evacuation Orders:** Delaying evacuation can endanger lives and complicate emergency response efforts. In anticipation of Hurricane Milton, Florida officials issued mandatory evacuation orders for vulnerable coastal areas. Despite warnings of "unsurvivable" storm surges, some residents chose to remain. Tampa Mayor Jane Castor issued a stark warning: "I can say this without any dramatization whatsoever: If you choose to stay in one of those evacuation areas, you are going to die."[1]

2. **Lack of Preparedness:** Failing to have a go bag or plan can lead to panic and delays.

3. **Underestimating the Severity:** Misjudging the severity of a disaster can lead to inadequate preparation and unnecessary risks.

4. **Poor Communication:** Failing to communicate your evacuation plan or location to family or authorities hinders coordination.

5. **Forgetting Essential Items:** Leaving behind crucial supplies like medications or documents complicates recovery efforts.

6. **Overreliance on GPS:** GPS can fail in areas with blocked signals or during network outages. Always have paper maps or local directions as backups.

7. **Ignoring Pet Safety:** Failing to prepare for your pet's evacuation needs can force you to leave them behind.

8. **Overloading Vehicles:** Packing too much in your vehicle can slow down evacuation and increase risks.

9. **Not Securing Your Home:** Forgetting to secure doors, windows, or utilities increases the chances of further damage.

10. **Returning Too Soon:** Prematurely reentering a disaster area can expose you to hazards and disrupt recovery efforts. Don't return home until local authorities declare it safe.

Emergency Preparedness Resources

- Emotional Recovery After a Disaster • https://www.redcross.org/get-help/disaster-relief-and-recovery-services/recovering-emotionally.html?

  Guidance on managing stress, recognizing common reactions to disasters, and strategies for emotional recovery.

- Disaster Preparedness Plan • https://www.redcross.org/get-help/how-to-prepare-for-emergencies/make-a-plan.html

  A step-by-step guide to creating an emergency preparedness plan, including identifying risks, developing communication plans, stocking supplies, and conducting evacuation drills.

- **Evacuation • https://www.ready.gov/evacuation**

  This is a planning guide for safely executing evacuations during emergencies, including routes, go-kits, and accommodations for disabilities and pets.

- **Preparing Your Family for Disasters • https://downloads.aap.org/AAP/PDF/disasters_family_readiness_kit.pdf**

  Emergency preparedness resources tailored for families with children include checklists, tips, and activities that prioritize children's well-being.

- **People with Disabilities • https://www.ready.gov/people-disabilities**

  Resources specifically designed to support emergency preparedness for people with disabilities.

CHAPTER 5

Training, Teamwork, and Caring

"True strength is knowing what to do, and doing it together." —Climate Cat

FIRST AID AND EMERGENCY RESPONSE TRAINING

Healthcare is shifting to an emergency footing as disasters and extreme weather events become the norm. Emergencies happen instantly. Help may take time to arrive, so equipping yourself with first aid and emergency skills could make you someone's lifeline.

Everyone can learn first aid basics, stepping into action when traditional systems are overwhelmed.

In emergencies, the individual often becomes the initial point of care rather than a clinic or hospital. One may need to stabilize injuries, call for assistance, and protect lives until professional help arrives. When resources are limited, individuals can play a critical role in supporting doctors, nurses, and first responders.

Emergency training enables decisive action in high-stress situations. First aid and Community Emergency Response Team (CERT) training equip individuals to build a circle of care and safety within their communities.

Community Emergency Response Team (CERT)

In 1985, Los Angeles officials visited Japan and observed how they trained neighborhood teams in specialized earthquake response skills. Later that year, a devastating earthquake in Mexico City prompted L.A. officials to recognize the need for structured community preparedness.

The Los Angeles Fire Department launched a pilot program in 1986 to train neighborhood leaders in fire suppression, search and rescue, and first aid. The program proved effective but lacked resources for expansion until the Whittier Narrows earthquake in 1987 galvanized public support.

In 1993, FEMA made the CERT program available to communities nationwide. CERT is designed to train volunteers in basic disaster response skills, including fire safety, light search and rescue, team organization, and disaster medical operations. The program is sponsored by local fire departments, emergency management agencies, state and local governments, and the Department of Homeland Security.

CERT trains you to assist your community during emergencies when professional responders may be delayed or overwhelmed. By participating in CERT, you become part of a coordinated local effort to prepare for, respond to, and recover from disasters, strengthening your town's or city's resilience. CERT programs emphasize community collaboration and preparedness, helping neighborhoods plan for emergencies. This builds a resilient support network that extends beyond the immediate crisis.

In CERT training, you'll learn:

- **Disaster Preparedness:** Understanding local hazards and developing personal and community emergency plans.

- **Fire Safety:** Identifying fire hazards, using fire extinguishers, and implementing basic fire suppression strategies.

- **Disaster Medical Operations:** Conducting patient assessments, controlling bleeding, treating shock, and performing basic first aid.

- **Light Search and Rescue:** Safely conduct search and rescue operations, including victim extrication and transport.

- **Disaster Psychology:** Recognize and address the psychological impact of disasters on survivors and responders.

EMERGENCIES HAPPEN INSTANTLY. HELP MAY TAKE TIME TO ARRIVE, SO EQUIPPING YOURSELF WITH FIRST AID AND EMERGENCY SKILLS COULD MAKE YOU SOMEONE'S LIFELINE.

CERT Training

State and local emergency management agencies sponsor CERT programs. They address regional issues, such as wildfire readiness in arid areas or flood response in coastal zones. You'll learn foundational skills for disaster response, such as hazard awareness, emergency medical care, and team organization.

Many fire and police departments partner with CERT programs and may host training sessions or provide information about how to enroll. I did my CERT training at a neighboring town's fire training school.

You can use the USA.gov State Emergency Management link to find your state's official emergency management agency. These agencies provide information on local risks, as well as alerts, evacuation routes, shelter locations, and preparedness resources: https://www.usa.gov/state-emergency-management.

First Aid

In the minutes before professional help arrives, a trained bystander can make the difference between life and death. You can act quickly by performing CPR, controlling bleeding, or stabilizing an injury.

Through first aid courses, you learn how to:

- Evaluate emergencies to ensure safety for both the responder and the victim.
- Restore breathing and circulation in unresponsive people, by performing CPR.
- Deliver controlled electric shocks to correct life-threatening heart rhythms, through the use of AEDs (automated external defibrillators).
- Apply techniques like direct pressure, dressings, and tourniquets to manage severe bleeding.
- Address burns, fractures, and sprains.
- Recognize and respond to heart attacks, strokes, and allergic reactions.

*Climate Cat Joins
the CERT Team*

Free Online First Aid Training

- **First Aid for Free** • https://www.firstaidforfree.com

  Offers online courses on first aid topics. Created by John Furst, author of *The Complete First Aid Pocket Guide*, it includes text-based lessons, quizzes, and certificates of completion.

- **Save a Life by NHCPS** • https://nhcps.com/lesson/cpr-first-aid-first-aid-basics/

  Provides CPR, AED, and First Aid certification training. Designed for healthcare professionals, teachers, fitness trainers, and caregivers, the program includes interactive materials, videos, and assessments.

- Alison • https://alison.com/course/basic-first-aid

  Offers self-paced first aid and CPR courses with step-by-step modules and provides certificates upon successful completion.

- Stop the Bleed • https://www.stopthebleed.org

  Offers training by the American College of Surgeons in controlling severe bleeding, including the use of direct pressure, tourniquets, and wound packing.

- The Mayo Clinic First Aid Guide • https://www.mayoclinic.org/first-aid

  Features a comprehensive library of first aid procedures, symptoms, and treatments, offering medically reviewed advice for various situations.

Enroll in an In-Person First Aid Course

For hands-on experience, in-person training is available at local hospitals, fire departments, EMS agencies, community colleges, adult education centers, YMCAs, and community centers. The American Red Cross, one of the country's largest volunteer organizations, offers certified first aid, CPR, and AED classes nationwide. You can use their online course finder to locate a class in your area: https://www.redcross.org/take-a-class/first-aid/first-aid-training.

BUILDING YOUR COMMUNITY NETWORK

A disaster, by definition, overwhelms resources. It causes such severe damage that state and local governments cannot respond without additional help. That help may come from you.

We have discussed the importance of family preparedness and the value of a solid emergency communication plan. We learned how first aid training and CERT programs can build personal skills and confidence.

We're taking the next step: working at the neighborhood level. This is about what you can do with the people living next door. You can create a local safety net to face challenges together.

The foundation of successful community preparedness begins with connections: friends, family, neighbors, and local organizations. According to FEMA's "Whole Community" approach to emergency management, effective disaster response begins at the neighborhood level. This creates "a secure and resilient nation with the capabilities required across the whole community to prevent, protect against, mitigate, respond to, and recover from the threats and hazards that pose the greatest risk."[1]

Communication

Plan a few reliable ways for you and your neighbors to stay in touch during emergencies, such as:

- ✓ Traditional phone calls for direct contact.
- ✓ Instant messaging on WhatsApp, Signal, Facebook Messenger, or Telegram for group chats.
- ✓ Two-way radios or other backup communication tools during power outages or network failures.
- ✓ Zello is a walkie-talkie-style app that allows real-time communication when phone networks are unreliable. Access it here: https://zello.com/

Choose a Point of Contact

Choose a neighborhood emergency coordinator or team to serve as a central point of contact. This person or group can help relay household information and connect with emergency services if needed. In larger neighborhoods, consider dividing the area into smaller zones, each with its own point of contact.

Apps such as Life360 (https://www.life360.com/) can help track the safety of individual family members, ensuring that everyone is accounted for during a crisis.

Building Community Expertise

Developing a resilient neighborhood requires more than just planning; it also requires specific skills. Encourage community members to pursue training in:

- Basic first aid and CPR certification
- Community Emergency Response Team (CERT) qualification
- Fire safety and basic search and rescue techniques
- Essential survival skills

Meeting Points

Establish safe, familiar places where neighbors can regroup during an emergency, just as you would for family meeting points. Select one nearby location for immediate needs and another location farther away for situations that require evacuation. Ensure that these meeting points are accessible and well known to all.

Plan for Specific Needs

Map out any special requirements within the neighborhood: households with older residents, young children, or people with disabilities. Ensure plans include support for pets and medical needs, and access to medications or equipment. Organize volunteers who can assist vulnerable neighbors during an emergency.

Document and Share the Plan

- ✓ Create a detailed contact list with the phone numbers, email addresses, and social media handles of all participating households.
- ✓ Evacuation routes specific to each area and designated emergency shelter locations.
- ✓ Lists of community resources, including food banks, shelters, churches, relief organizations, and community centers.
- ✓ Inventory of shared physical resources, like generators or emergency supplies.
- ✓ Special considerations for people who require some extra help.
- ✓ Distribute the guide digitally and in print to all households, and ensure it's updated regularly.

Practice

Host community drills and training sessions to ensure everyone knows their role in an emergency. Practice using communication tools, locating meeting points, and responding to scenarios like a wildfire or flood. Pressure test it. Ask questions: What could go wrong? What happens *if?* Regular rehearsal helps build familiarity with the plan.

Stay Updated

Follow official emergency management sources such as Ready.gov and subscribe to local alert systems for real-time alerts and updates. Monitor weather forecasts and disaster reports, and share *verified* news and essential information within your network.

Food Banks

Food banks serve as lifelines in our communities, supporting individuals and families facing food insecurity. Feeding America operates the nation's largest network of food banks, working through local partners to provide emergency food assistance to millions. Their network includes more than 200 food banks and 60,000 food pantries and meal programs. They work to distribute food while preserving dignity and respect for everyone they serve, recognizing that anyone might need assistance at some point.

To find a food bank near you, visit Feeding America's local food bank locator at https://www.feedingamerica.org/find-your-local-foodbank. The USDA's Local Food Portal connects communities with food resources and programs, including food banks, school meals, and nutrition assistance. It helps coordinate federal aid with local organizations to ensure food reaches those who need it most. For more information, visit https://www.usdalocalfoodportal.com/.

EMERGENCY PREPAREDNESS FOR PETS

Be sure to include your pets in your climate emergency preparations, from putting together a pet-specific emergency kit to planning how you'll evacuate together. As extreme weather events increase, so do the risks to our pets. Pets depend on us, so our emergency plans need to include them.

A study from the ASPCA found that while 90 percent of pet owners said they would evacuate with their pets, only 46 percent have actual preparedness plans. "More than 1 in 5 pet owners have evacuated their homes due to a disaster or emergency and nearly half left at least one pet behind when they evacuated."[1]

Abandoning Pets: Leaving pets behind can cause the pets enormous distress, trauma, and even death. Abandoning pets during extreme weather may even lead to legal consequences. Penalties include fines, community service, or imprisonment. Owners can be held liable for property damage or public safety threats caused by abandoned pets.

People often refuse to evacuate in emergencies because they do not want to leave their pets behind. After Hurricane Katrina, the Pets Evacuation and Transportation Standards (PETS) Act of 2006 was passed. The law requires state and local emergency preparedness plans to account for the needs of individuals with household pets and service animals during major disasters or emergencies. It ensures that pets are included in evacuation, rescue, and sheltering plans, recognizing that many people will refuse to evacuate without their animals.

FEMA, the Red Cross, and CERT programs work with states and local governments to develop pet-friendly shelters, transportation plans, and emergency services for people and their animals.

Understanding Risks to Your Pets' Health

If heat affects you, it can and will affect your pets. Pets overheat more quickly than we do, especially those with thick fur. Pets with certain health conditions are more prone to overheating due to their body's inability to regulate temperature effectively.

Brachycephalic Syndrome: This condition is common in flat-faced breeds such as bulldogs, pugs, and Persian cats. These breeds have shorter airways, making it harder to cool down by panting.

Heart Disease: Pets with heart conditions struggle with circulation and oxygen delivery, impairing their ability to dissipate heat.

Respiratory Disorders: Conditions like laryngeal paralysis or chronic bronchitis reduce adequate airflow, limiting their ability to cool themselves.

Obesity: Overweight pets have extra fat (which acts as insulation), making it harder for them to cool down. Excess weight can strain their cardiovascular and respiratory systems.

Diabetes and Endocrine Disorders: Conditions like Cushing's disease or hypothyroidism can alter metabolic function and heat regulation.

Neurological Conditions: Pets with neurological issues may have impaired signals to regulate body temperature.

Arthritis or Mobility Issues: Pets with reduced mobility may be unable to move to cooler areas or position themselves to cool off effectively.

Skin Disorders: Skin infections or irritation can impair heat dissipation through the skin.

Age: Old and very young pets often have more trouble regulating their temperature.

Medication Side Effects: Drugs, such as antihistamines, sedatives, or diuretics, can affect your pet's ability to handle heat.

AS EXTREME
WEATHER EVENTS
INCREASE, SO
DO THE RISKS TO
OUR PETS. PETS
DEPEND ON US, SO
OUR EMERGENCY
PLANS NEED TO
INCLUDE THEM.

Keeping Pets Cool During a Heat Wave

Plan for emergencies by preparing ways to keep your pet cool during power outages, such as using battery-powered fans and setting up cool resting spots. Ensure you have a pet first aid kit on hand and know the signs of severe heat exhaustion. Always provide fresh water for your pets by placing extra bowls around your home and adding ice cubes to help keep the water cool.

Cooling strategies:

✓ Place frozen water bottles wrapped in towels in your pet's bed or resting area. Use cooling mats or wet towels for pets to lie on.

✓ Use pet-specific cooling products like cooling vests, collars, and mats.

✓ You can tie wet bandanas around your pet's neck for added cooling.

✓ Offer pets frozen treats like ice cubes, frozen fruits, or specially-made frozen pet treats to help them cool down.

✓ Use fans or air conditioning to keep your home cool.

✓ Allow pets to rest in the coolest part of the house.

Limit Outdoor Time

Keep pets indoors during the hottest parts of the day. Walk dogs early in the morning or late in the evening when temperatures are cooler. If pets must be outside, ensure they have access to shaded areas and proper ventilation. Use umbrellas, canopies, or shade sails to create cool spots.

Grooming

Keep pets well groomed to help regulate their body temperature. For pets with thick fur, consider trimming their coat.

Avoid Hot Surfaces

Protect your pet's paws from hot pavement or asphalt. Walk them on grass or shaded paths, and check the temperature of surfaces before allowing them to walk on them.

Monitor for Heat Stress

Watch for signs of heat stress, such as excessive panting, drooling, lethargy, and vomiting. If you notice these signs, immediately move your pet to a cooler area and contact your veterinarian.

Risks Pets Face During Extreme Events

Let's review some specific risks pets face during power outages, floods, and hurricanes. These extreme events can disrupt access to food and water, exposing pets to physical harm or illness.

Power Outages

Loss of heating or cooling makes it hard for pets to stay warm in winter or cool in summer. Small pets, such as birds, hamsters, or reptiles, require specific temperature and humidity ranges; extreme conditions can quickly become deadly.

If the power is out for extended periods, pets who rely on refrigerated diets or medicines will also face health risks. Pets unfamiliar with dark spaces may become stressed or injured while moving around. When the power goes out, automatic feeders, water dispensers, and aquarium filtration systems will fail.

Floods

Pets left behind or unable to escape rising waters may drown. Floodwaters can expose pets to bacteria, chemicals, and other toxins, causing gastrointestinal or skin issues. Standing water increases the risk of leptospirosis and other waterborne diseases. Floods may separate pets from their owners or lead to injuries.

Hurricanes and Tornadoes

Extreme weather can cause pets to panic and run, increasing the risk of injury or loss. Pets can be struck and injured by debris during high winds. Flooding and storm surges increase the risk of exposure to toxic floodwaters and drowning.

Loud winds, thunder, and atmospheric pressure changes can cause extreme anxiety. Evacuations can be chaotic, and pets without proper identification can become lost and separated from their owners. Tornadoes destroy kennels, homes, and other safe spaces, and stir up dust, chemicals, and debris that are harmful to pets.

Basic Preparation

✓ Ensure your pets are up to date on vaccinations and receive basic obedience training.

✓ Use tags, microchips, and collars with identification markers to ensure your pets can be easily identified and returned if they get lost.

✓ Understand how your pets react to stress. Cats may hide, while dogs may become more protective or aggressive.

✓ Acclimate your pets to carriers. Practice using carriers and traveling in your vehicle to reduce stress during an evacuation.

✓ Have a photo of you and your pets on your phone for identification.

Build an Emergency Evacuation Kit for Your Pet

For the most part, everything you might need (food, medicine, water, sanitation, comfort), your pets need as well:

✓ Store enough food and water in airtight, waterproof containers for several days. Also, remember food and water bowls, dish soap, and a manual can opener.

✓ Pack a two-week supply of medicines and first aid supplies, including prescription medications and disinfectant, in a waterproof container.

✓ Include sanitation and comfort items such as litter boxes, litter, waste bags, blankets, and absorbent pads.

✓ Have your pets' medical records and recent photos on hand. It's best to have pictures of you and your pet together.

✓ Pack a backup collar with ID tag, harness or leash, and copies of your pet's registration information.

✓ Bring familiar items like toys, treats, or bedding to reduce stress.

✓ Include a flashlight and blankets in your packing.

Evacuation planning should include safe routes and pet-friendly destinations. Identify shelters or hotels that accommodate pets, and use resources such as PetsWelcome.com (https://www.petswelcome.com/) to locate suitable options along your route. As soon as you become aware of a potential disaster, consider booking a cancellable hotel that allows pets to ensure you have a secure place for them if evacuation becomes necessary.

Under the PETS Act, jurisdictions that seek FEMA assistance must plan for household pets and establish pet-friendly sheltering options, but not every human shelter accepts pets; service animals, however, are always permitted. Many communities offer pet-friendly evacuation centers designed to keep families and their animals together.

Here are the things you will need to bring with you to an emergency shelter:

✓ Town license

✓ Copies of veterinary records

✓ Proof of ownership

✓ Rabies tag(s)

✓ A photo of you with your pet

✓ An emergency contact list, including neighbors and your veterinarian

✓ Any medications your pet needs

✓ If your pet is microchipped, ensure your contact information is current

If local shelters are full or unable to accommodate pets, emergency management agencies may coordinate with nearby kennels, rescues, or foster networks. Community resources can also play a key role. Neighbors, friends, and family members should work together to create a support network for pet care. A buddy system can ensure that someone is available to care for or evacuate your pets if you are unable to do so yourself.

Some emergency management agencies partner with animal control or rescue organizations to provide transportation during evacuations. They may offer services such as microchipping clinics and vaccination drives to help prepare pets for emergencies, and they often maintain lists of local veterinarians who can provide support.

In many regions, specialized animal response teams such as County Animal Response Teams (CART) and Disaster Animal Response Teams (DART) are trained to assist in evacuating pets and livestock. CARTs operate at the county level and are responsible for planning, coordinating, and carrying out animal-related disaster response efforts within a specific county.

These responsibilities include setting up temporary pet shelters, rescuing stranded animals, and assisting with livestock evacuations in coordination with local emergency management agencies. DARTs function at the regional, state, or national levels and provide support in large-scale disasters. They assist with the rescue, sheltering, and care of animals affected by events such as hurricanes, wildfires, or floods. These teams typically consist of trained volunteers with experience in both animal care and emergency response.

Pet First Aid

The American Red Cross offers an online course on first aid skills for cats and dogs.[2] This training covers:

- Understanding and assessing your pet's vital signs
- Breathing and cardiac emergencies
- Wounds and bleeding
- Seizures
- Preventative care

The Red Cross offers a Pet First Aid app with instructions for first aid and emergencies, including treating wounds, controlling bleeding, and caring for breathing and cardiac emergencies.[3] After a disaster, it's vital to help pets adjust by being patient and observant of any behavioral changes they may exhibit.

Establishing a routine and familiar environment can provide stability. Before allowing pets to roam freely, inspect your home and surroundings for potential hazards that could harm them. Monitor your pets for signs of stress or health issues, and consult your veterinarian if any concerns persist.

Special Considerations for Birds

Ensure proper ventilation and temperature control.

Have a misting bottle and fabric cage cover for smoke or air quality issues.

Train your bird to return to its cage or a safe spot.

Have a portable heating source for colder conditions.

SECONDARY EFFECTS OF EXTREME WEATHER

CHAPTER 6

Insects and Disease

"Ticks and mosquitoes are the new cartographers of climate." —Climate Cat

BITES: EXTREME WEATHER AND INSECTS

I love monarch butterflies and their vivid black-and-orange wings. Monarch caterpillars, with their soft, silky, striped bodies, as they munch on milkweed. Dragonflies, with a helmet of panoptic eyes, which see in all directions. Their metallic colors look like an Art Nouveau Tiffany lamp. Spiders rebuild their dew-frosted webs each morning and test the tensile strength of their geometric silk.

This chapter focuses on bugs I do not love.

I'm going to talk about skeeters. I'm going to talk about ticks. I'm going to talk about bug spray. We often think of bugs as just an irritation. But people with severe Lyme disease can tell you that bug bites can destroy your life.

What does extreme weather have to do with bugs? More extreme weather means more bugs, which means more disease. Climate change is, at its core, nature change. As environments transform, water levels fluctuate, and seasonal patterns shift, larger populations of disease-carrying insects are emerging in new locations.

An insect-borne disease is an illness spread to humans when infected insects bite us. Examples include mosquito-spread diseases like malaria (a parasite) and West Nile (a virus), and tick-spread diseases like Lyme disease (a bacterium).

What are these diseases, and what insects are spreading them?

DISEASE	DESCRIPTION	BUG	ESTIMATED ANNUAL U.S. CASES
Lyme disease	A bacterial infection sometimes causing a distinctive bull's-eye rash that can lead to joint, heart, and neurological problems if left untreated.	Tick	500,000–600,000
Alpha-gal syndrome (AGS)	An allergic reaction to red meat, triggered by tick bites, that causes hives, digestive issues, and potentially anaphylaxis (shock).	Tick	15,000 (the actual number of AGS cases in the United States is not known, but as many as 450,000 people may be affected)
Anaplasmosis	A bacterial infection causing fever, headache, muscle pain, and reduced white blood cell and platelet counts.	Tick	5,655
Rocky Mountain spotted fever	A potentially fatal bacterial disease causing fever, headache, and a characteristic spotted rash.	Tick	5,544
West Nile virus	It is the most common mosquito-borne disease in the U.S. It causes flu-like symptoms that can progress to serious neurological illness.	Mosquito	2,647 were reported, but many more people were assumed to be infected with no symptoms.
Babesiosis	A parasitic infection similar to malaria that destroys red blood cells and can cause anemia.	Tick	2,368
Ehrlichiosis	A bacterial disease with flu-like symptoms that can become severe if not treated promptly.	Tick	2,093

MORE EXTREME
WEATHER MEANS
MORE BUGS,
WHICH MEANS
MORE DISEASE.

The number of reported cases of these diseases is often much lower than the actual number of infections. Many people do not show symptoms or experience only mild illness and never get tested. For example, about eight out of ten people infected with West Nile virus have no symptoms.

Diseases like Zika virus can cause very mild symptoms, or none at all, making them easy to overlook. Even Lyme disease can go undiagnosed for months if the characteristic rash does not appear. As a result, public health reports often underestimate the number of people infected.

Why Are There So Many Mosquitoes and Ticks?

1. Warmer temperatures accelerate insect life cycles, increasing populations and expanding their geographical range. Prolonged warm seasons and increased humidity extend breeding periods and enhance larval survival.

2. Changes in rainfall patterns and shifts in wind direction create new breeding grounds for insects, further expanding their range and the spread of insect-borne diseases.

3. Abundant acorns and milder winters boost small-mammal hosts, which in turn fuel tick booms.

4. Urban heat islands create favorable breeding conditions, increasing diseases in densely populated cities. With their heat and stagnant water, urban areas are ideal mosquito breeding grounds.

Mosquitoes

Mosquitoes are more obvious than ticks. We see, feel, and hear them, where ticks can be so small they're never detected. Mosquitoes carry a laundry list of diseases: dengue, Zika, chikungunya, yellow fever, malaria, and West Nile virus, spread by four main species:

MOSQUITO SPECIES	APPEARANCE	BEHAVIOR	DISEASES TRANSMITTED	GEOGRAPHIC DISTRIBUTION
Aedes aegypti (yellow fever mosquito)	Dark brown to black with white lyre-shaped markings on the thorax and white-banded legs	Aggressive daytime feeders that prefer humans as their meal; they breed in small water containers near human habitations.	Dengue, Zika, chikungunya, yellow fever	Southern U.S. (Florida, Texas, Gulf Coast)
Aedes albopictus (Asian tiger mosquito)	Black with white stripes on the body and legs; single white stripe down the center of the thorax	Less aggressive; they breed in various water containers; adaptable to cooler temperatures.	Dengue, Zika, chikungunya	Temperate regions in the U.S.
Anopheles (malaria mosquito)	Generally brown or black with sensory organs as long as the feeding organ	Night feeders that prefer humans; they breed in various aquatic habitats.	Malaria, lymphatic filariasis	Southern U.S.
Culex (house mosquito)	Generally brown or gray	Active at dawn, dusk, and night; they breed in stagnant water.	West Nile virus, St. Louis encephalitis, Western equine encephalitis	Widely distributed across the U.S.

West Nile Virus

West Nile virus is the most common mosquito-borne disease in the continental U.S. It's primarily spread by the Culex (common house) mosquito, which becomes infected after biting birds that carry the virus. Symptoms range from mild flu-like symptoms to severe neurological conditions and even death. Droughts play a major role in West Nile outbreaks. During droughts, water bodies shrink, concentrating birds (the primary hosts of West Nile virus) and mosquitoes.

How to Be Aware

West Nile virus is typically active from late spring to early fall, with peak activity in summer. Stay aware by:

- ✓ Checking your local or state health department website for alerts or updates.
- ✓ Listening to news reports or public health announcements during mosquito season.
- ✓ Watching for mosquito control signs in your neighborhood, such as fogging or spraying notices.
- ✓ Paying attention to reports of dead birds, especially those of crows or jays, which are common carriers.

How to Track the Spread

- ✓ State health departments track and publish data on West Nile activity, including positive mosquito pools, infected birds, and human cases. A "mosquito pool" is a batch of mosquitoes (typically ten to fifty) collected from traps in a specific area and tested together as a single sample.
- ✓ Websites like the CDC's ArboNET provide maps and weekly updates. (https://www.cdc.gov/mosquitoes/php/arbonet/index.html)
- ✓ Local mosquito control districts often share real-time data on mosquito traps, surveillance efforts, and spraying schedules.
- ✓ You can report dead birds to local authorities to help track potential outbreaks.

Bug Spray Recommendations

- ✓ Use EPA-approved insect repellents such as DEET, picaridin, and oil of lemon eucalyptus (OLE).
- ✓ Apply repellents before going outside, especially at dawn and dusk when Culex mosquitoes are most active.

✓ Wear long sleeves and pants when outdoors in the evening or early morning.

✓ Clean gutters, empty birdbaths, pet bowls, and flower pot saucers to eliminate standing water around your home.

✓ Use window and door screens and sleep under mosquito nets if needed.

Symptoms of West Nile Virus

Most people (approximately 8 in 10) infected with West Nile virus do not experience any symptoms. However, when symptoms occur, they can range from mild to severe:

Mild Symptoms: fever, headache, body aches, fatigue, skin rash (occasionally).

Severe Symptoms (Neuroinvasive Disease): high fever, neck stiffness, disorientation, tremors or seizures, muscle weakness or paralysis, coma or death in rare cases.

- Severe cases are more likely in older adults and those with weakened immune systems.

What to Do If You Get Infected

✓ There is no specific antiviral treatment for West Nile virus.

✓ Rest, stay hydrated, and use over-the-counter pain relievers like acetaminophen or ibuprofen for mild symptoms.

✓ If you have severe symptoms, seek medical attention immediately. Hospitalization may be required for intravenous fluids, pain management, and neurological support.

✓ If you are diagnosed, notify your local health department so they can take steps to reduce the spread and notify others in your area.

Diseases You Can Bring Home from Travel

Traveling outside the continental U.S. can expose you to mosquito-borne diseases. Dengue virus transmission remains high in U.S. territories like Puerto Rico and the U.S. Virgin Islands. Here's how to recognize these diseases, reduce risk, and know what steps to take.

Symptoms of Mosquito-borne Illnesses

Be vigilant about symptoms after returning home:

> **Malaria:** Fever, chills, sweats, headache, muscle aches, fatigue. Symptoms may appear days or even weeks after returning.
>
> **Dengue:** Sudden high fever, severe headache, pain behind the eyes, joint and muscle pain, rash, mild bleeding.
>
> **Chikungunya:** Fever, severe joint pain (often debilitating), headache, muscle pain, rash.
>
> **Zika:** Fever, rash, joint pain, conjunctivitis (red eyes). Note: this disease poses severe risks during pregnancy.

Awareness

Before traveling, check the CDC Travel Health Notices website for alerts about disease risks or outbreaks at your destination. Physicians in the U.S. are advised to consider dengue in any patient who develops symptoms after recent travel to affected regions.

How to Stay Informed

✓ The CDC Travel Health Notices are regularly updated with details about disease outbreaks and high-risk areas. (https://www.cdc.gov/mosquitoes/prevention/preventing-mosquito-bites-while-traveling.html)

✓ Review the CDC's Guide on Preventing Mosquito Bites While Traveling for practical, detailed advice on staying safe abroad.

Prevention Tips

✓ Use EPA-approved insect repellents such as DEET, picaridin, and oil of lemon eucalyptus (OLE).

✓ Wear long-sleeved shirts and long pants to minimize exposed skin.

✓ Stay in accommodations with screens or air conditioning; use mosquito nets when necessary.

✓ If you are traveling to malaria-risk areas, take antimalarial medications as recommended by your doctor. There are also vaccines for yellow fever and chikungunya.

What to Do If You Become Sick

Seek immediate medical care if you experience fever or the symptoms listed above after returning from travel. Inform your doctor of your recent travel history. Early diagnosis and prompt treatment can significantly reduce the severity of these diseases.

Climate Change and the Increased Spread in the U.S.

Rising temperatures and altered rainfall patterns have expanded the range of Aedes mosquitoes, raising concerns about new outbreaks of diseases like dengue and Zika within the continental U.S. Aedes mosquitoes are projected to spread as far north as Chicago by 2050. Recent localized outbreaks have already occurred in states like Florida and Texas.

How to Stay Aware

✓ Regularly check your state or local health department websites for alerts, mosquito surveillance data, and updates on mosquito-borne diseases. (https://www.cdc.gov/han/php/about/index.html)

✓ Subscribe to public health alert systems, such as the CDC's Health Alert Network (HAN), for real-time disease outbreak notifications.

✓ Follow local news and community advisories, especially during warmer months when mosquito activity peaks.

✓ You can participate in community mosquito surveillance efforts, which involve tracking mosquito traps, testing for pathogens, and reporting dead birds or clusters of illness.

TICKS

Ticks are tiny arachnids that thrive in warm, humid environments. As global temperatures rise, tick populations expand into new regions, extending their active seasons and increasing the exposure of more people to health risks. Warm weather and high humidity give ticks more days each year to actively feed and reproduce. Black-legged ticks are now found in parts of Canada that previously were too cold for their survival.

The animals that ticks feed on (deer, mice, and birds) have altered their migration patterns and expanded their ranges. This movement of host animals carries ticks into new territories, exposing previously unaffected human populations to tick-borne diseases. Habitat fragmentation, such as suburbs built near wooded areas, creates ideal environments for ticks and their hosts, including deer and mice.

Many regions have experienced significant increases in deer populations, providing more hosts for ticks to feed on and multiply. Reduced biodiversity benefits mice, which thrive when their natural predators decline, boosting tick populations. As they migrate, birds can carry ticks over long distances, helping ticks spread into new territories.

Tick Life Cycle

Ticks have a four-stage life cycle: egg, larva, nymph, and adult. Unlike many insects that complete their life cycle in weeks or months, ticks can take up to three years to progress from egg to adult, depending on environmental conditions and host availability.

The life cycle begins when an adult female tick, after feeding and mating, lays thousands of eggs in sheltered locations like leaf litter or thick grass. These eggs hatch into larvae, often called "seed ticks," which begin searching for their first meal (blood). Larvae emerge uninfected with pathogens; they acquire them only after feeding on infected hosts, typically small mammals such as mice or birds.

The nymph stage is the most dangerous to humans. At this point, ticks are the size of a poppy seed, making them incredibly difficult to spot. Their small size and ability to transmit pathogens acquired during their larval stage make nymphs responsible for most human tick-borne infections. While more visible, adults still pose significant risks. Their larger size makes them easier to detect, but their bites can transmit serious diseases, particularly if they remain attached for extended periods.

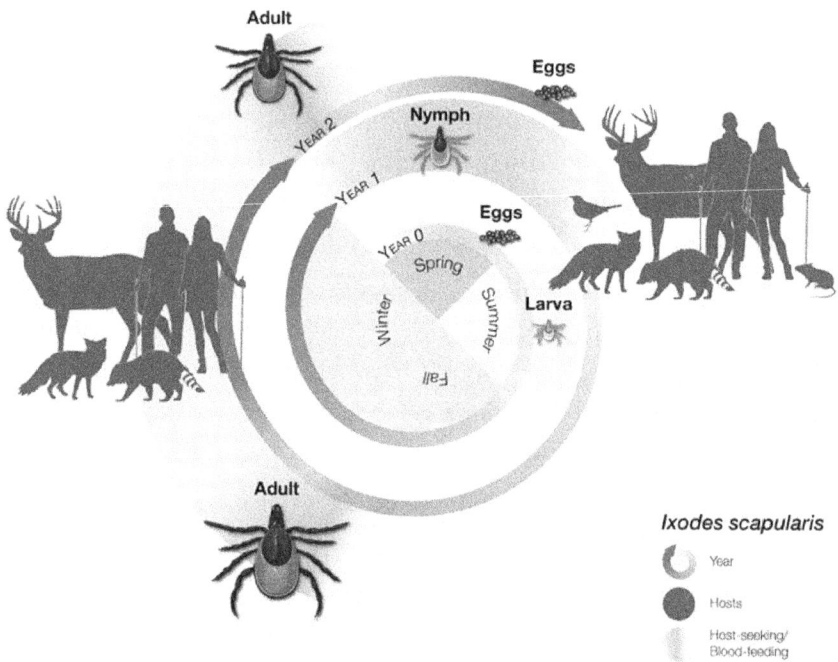

Source: Centers for Disease Control and Prevention, Tick Lifecycles, CDC.gov, https://www.cdc.gov/ticks/about/tick-lifecycles.html.

Lyme Disease

I want to focus on Lyme disease because it's rapidly spreading across the country into regions where it was previously unknown.

This isn't just abstract data to me. I've watched my friends and family suffer serious consequences like facial paralysis from untreated infections. What makes Lyme so dangerous is how frequently it's misdiagnosed, even by experienced healthcare providers, because symptoms can mimic so many other health conditions.

People may not connect their symptoms to a tick, and they may not have noticed the bite in the first place since the "bull's-eye rash" doesn't always appear. By raising awareness about tick prevention, symptoms, and the importance of early treatment, I aim to help you avoid long-term health impacts.

The black-legged tick, which transmits Lyme disease, is prevalent in the Northeast and Upper Midwest but is spreading throughout the country. Estimates are as high as 630,000 Lyme cases annually. Due to the tick's lifecycle, the most common months for developing Lyme disease are June and July.

The highest rates of Lyme disease are observed among children aged 5 to 9 and adults aged 50 to 79.

Source: Centers for Disease Control and Prevention, Public Health Image Library, image ID 1669, CDC, https://phil.cdc.gov/Details.aspx?pid=1669.

WHAT MAKES LYME
SO DANGEROUS IS
HOW FREQUENTLY
IT'S MISDIAGNOSED,
EVEN BY EXPERIENCED
HEALTHCARE PROVIDERS,
BECAUSE SYMPTOMS
CAN MIMIC SO
MANY OTHER HEALTH
CONDITIONS.

The black-legged tick (deer tick) spreads more than just Lyme disease. It can also transmit *anaplasmosis*, which causes flu-like symptoms. This includes fever, headaches, muscle pain, and extreme fatigue, making daily activities difficult. Many people don't realize they have anaplasmosis because the symptoms are similar to those of other common illnesses. The good news is that when caught early, doctors can usually treat it effectively.

Lyme Symptoms and Progression

Early Symptoms: Fever, chills, headache, fatigue, muscle and joint aches, bull's-eye rash.

Later Symptoms (If Untreated): Severe headaches and neck stiffness; additional rashes on other areas of the body; arthritis with severe joint pain and swelling, particularly in the knees and other large joints; facial palsy/Bell's palsy (loss of muscle tone or droop on one or both sides of the face); intermittent pain in tendons, muscles, joints, and bones; heart palpitations or irregular heartbeat (Lyme carditis); dizziness or shortness of breath; inflammation of the brain and spinal cord; shooting pains, numbness, or tingling in the hands or feet.

Lyme Disease is Frequently Misdiagnosed

Some of the common misdiagnoses are:

Flu: Due to fever, chills, headache, and fatigue.

Chronic Fatigue Syndrome: Because of persistent fatigue and malaise.

Fibromyalgia: Joint and muscle pain can be confused with this condition.

Rheumatoid Arthritis: Severe joint pain and swelling can mimic rheumatoid arthritis.

Multiple Sclerosis (MS): Neurological symptoms like facial palsy and nerve pain can resemble MS.

Depression or Anxiety Disorders: Mood changes and fatigue might be attributed to psychological causes.

Meningitis: Severe headaches and neck stiffness can lead to a misdiagnosis of meningitis.

Heart Disease: Palpitations or irregular heartbeats might be misinterpreted as heart disease.

How to Talk to Your Doctor About Lyme

Early diagnosis and treatment are crucial to prevent long-term complications. Antibiotics like doxycycline or amoxicillin are the standard treatments. For help communicating with your doctors and increasing the chances of an *accurate* Lyme disease diagnosis, here are some recommendations:

Document Symptoms: Record all symptoms, including fever, fatigue, muscle pain, and joint stiffness. Note their onset, severity, and progression. Take dated photos from multiple angles for rashes, particularly the "bull's-eye" pattern. Just be aware that not all tick bite cases develop this distinctive rash.[1]

Background: Tell your doctor about recent exposure to wooded or grassy areas, especially in regions with a known tick presence. Mention any potential bug bites, even if they are minor.

Ask for Both ELISA and Western Blot Tests: These are the most reliable diagnostic tools for Lyme disease. Consider retesting if initial results are inconclusive because antibodies may take weeks to develop. Discuss testing for co-infections like Anaplasma or Babesia if symptoms are severe or unusual.

Discuss Overlapping Conditions: Since Lyme symptoms can mirror other conditions like chronic fatigue syndrome or rheumatoid arthritis, share any relevant medical history. Ask your doctor to evaluate Lyme alongside these other conditions.

Be Persistent: If symptoms persist despite negative tests, request follow-up testing and monitoring. If symptoms strongly align with Lyme disease patterns, diagnosis can be based on symptoms alone. For ongoing symptoms post-treatment, discuss the possibility of chronic Lyme disease or post-treatment Lyme disease syndrome (PTLDS).

Emerging Tick-Borne Diseases

While I was writing this book, I learned about someone going into shock in a restaurant near me. After several tests, the doctors determined it was alpha-gal syndrome (AGS), an allergy to red meat caused by tick bites. It occurs when a sugar molecule called alpha-gal, found in tick saliva, enters the bloodstream and triggers an immune response.

In the U.S., AGS is linked to the lone star tick, and there are about 15,000 new positive test results for AGS annually. (The actual number of AGS cases in the United States is not known, but as many as 450,000 people may be affected.) Symptoms typically occur 2–6 hours after consuming red meat or dairy, but can also appear immediately after exposure to other products containing alpha-gal. These are medical products and medications (like the cancer drug cetuximab) and products made from mammals (gelatin and personal care items).

Symptoms range from mild to severe and can be life-threatening. Skin reactions include itchy skin, hives, and swelling. Gastrointestinal issues include abdominal pain, diarrhea, heartburn, and nausea. Lung symptoms include anaphylaxis, cough, and shortness of breath. Other symptoms include heart palpitations, low blood pressure, and arthritis.

Diagnosis involves reviewing symptoms and health history, as well as conducting allergy tests, such as blood tests for alpha-gal antibodies and skin prick tests. Treatment includes avoiding foods and products with alpha-gal and using medications like antihistamines or epinephrine injectors for accidental exposure.

Tick Protection

Clothing

✓ Wear long-sleeved shirts and pants, preferably in light colors, to make ticks easier to spot. Light-colored clothing helps you notice and remove ticks before they attach.

✓ Tuck pants into socks. This creates a physical barrier against climbing ticks and reduces their ability to access your skin.

✓ Use a lint roller on your skin or clothing after you come indoors to pick up any wandering ticks before they attach.

✓ Use permethrin-treated clothing. Permethrin is a powerful tick repellent that bonds to fabric, offering long-lasting protection through multiple washes. Clothing pre-treated with permethrin is also commercially available.

Tick Checks

After every outdoor activity, systematically inspect yourself, your children, and your pets. Check behind and inside the ears, along the hairline and scalp, in the underarms and between the fingers, inside the belly button, around the waist and beltline, between the legs and behind the knees, at the sock line, and between the toes. Tick check routine:

Shower Promptly: Showering within two hours of coming indoors can help wash off unattached ticks and provide an opportunity for a thorough tick check.

Dry Clothes: Tumble-dry clothes on high heat for 10 minutes if they're already dry, and up to 60 minutes if damp, to kill any ticks.

Use Tools: A magnifying glass can help spot tiny nymph ticks that resemble specks of dirt.

Check Pets Carefully: Ticks will latch onto pets, which can then bring them into your home. Use a fine-tooth comb to inspect your pet's fur, especially around the neck, ears, and belly.

*Climate Cat Keeping a Close Eye
on Creepy Crawlers*

Tick Removal and Testing

If you find an attached tick, remove it quickly and correctly to minimize the risk of disease transmission.

Use fine-tipped tweezers. Grasp the tick as close to the skin's surface as possible.

Pull upward steadily and evenly. Avoid twisting or jerking, as this can cause the mouthparts to break off and become embedded in the skin.

Clean thoroughly. Wash the bite area and your hands with soap and water, then apply an antiseptic.

To kill the tick safely, seal it in clear tape or place it in a sealed container with a slightly moistened paper towel.

Preserve the tick in a sealed container or plastic bag in case symptoms develop. Testing the tick can identify its species and detect pathogens, which may guide medical treatment. Testing labs include tickcheck.com, tickreport.com, and ticknology.org. Early recognition of symptoms is key to effective treatment.

Watch for fever, fatigue, joint pain, or the characteristic bull's-eye rash of Lyme disease. If any symptoms appear, seek medical attention promptly. The CDC's Tick Bite Bot offers guidance on assessing risk and determining when to consult a healthcare provider. Saving the tick can help medical professionals evaluate potential disease exposure.

Tick Guidance for Pets

Ticks attach near pets' heads, necks, ears, and feet. Monitor pets for visible ticks (initially pinhead-sized, then growing larger as they fill with blood) and check them regularly after outdoor activities.

Insect and Tick Monitoring

- TickEncounter Resource Center • https://web.uri.edu/tickencounter

  Provides risk maps for tick activity and exposure, and a program where individuals can submit photos of ticks for identification and feedback.

- Bay Area Lyme Foundation • https://www.bayarealyme.org/

  Information on Lyme disease prevention, symptoms, and treatment, plus funding for scientific research.

- Global Lyme Alliance • https://www.globallymealliance.org/

  Resources for patients, caregivers, and healthcare professionals, including guides to symptoms and treatments.

- CDC Tick Bite Bot • https://www.cdc.gov/ticks/communication-resources/tick-bite-bot-1.html

  An interactive tool providing step-by-step guidance for tick bite prevention and response.

BUG SPRAYS

Let's talk about keeping bugs at bay. In this chapter, I'll walk you through the world of insect repellents and insecticides: what works, how they work, and the smartest ways to use them to protect yourself.

Insects, particularly mosquitoes, locate humans by detecting chemicals like carbon dioxide, lactic acid, and octanol emitted from our skin and breath. Bug sprays work through different mechanisms to protect us:

Scent-blocking repellents, such as DEET, picaridin, and OLE, confuse and overwhelm insects' olfactory receptors, making it difficult for them to find us. Insecticides, such as permethrin, kill insects upon contact rather than just repelling them.

Types of Bug Spray

DEET:

- (N, N-diethyl-meta-toluamide)
- Developed by the U.S. military in the 1940s, has been a staple in insect protection for decades
- Products containing 15 to 30 percent DEET protect for up to 8 hours
- Effective against mosquitoes, ticks, and flies
- EPA studies have found no evidence that DEET causes cancer when used correctly
- Concentrations above 30 percent are not recommended
- It may cause skin irritation and, in rare cases, neurological effects

Picaridin:

- A newer and highly effective alternative to DEET
- Introduced to the U.S. market in 2005

- Similar in chemical structure to compounds found in black pepper plants
- 20 percent picaridin formulations show exceptional long-lasting protection
- Safe for pregnant women and children over two months old
- Less likely to irritate the skin than DEET

Oil of Lemon Eucalyptus:

- (OLE)
- A plant-based alternative with proven effectiveness
- Derived from the lemon eucalyptus tree
- Contains p-menthane-3,8-diol (PMD), which provides repellent properties
- Repellents with 30 percent OLE are effective for up to 7 hours
- It may cause eye irritation
- Not recommended for children under three years old

Permethrin:

- Apply to clothing, shoes, socks, tents, backpacks, sleeping bags, and other fabric items
- Once applied and thoroughly dried (typically after 24 hours), it bonds to fabric fibers
- Remains effective through multiple washes
- Protects against mosquitoes, ticks, flies, and other biting insects

Permethrin Safety Considerations

- EPA-approved for use on clothing and gear
- Generally safe for humans once dry
- Toxic to cats if they come into contact with wet, freshly treated fabrics. Keep wet or freshly treated items away from cats until completely dry

- Products specially formulated for dogs are safe when used according to instructions

For more guidance on finding repellents, check: https://www.epa.gov/insect-repellents/find-repellent-right-you

Best Practices for Applying Repellents

Skin Application (for DEET, Picaridin, and OLE)

Apply a thin film to exposed skin (look for a slight shimmer). Never apply directly under clothing; instead, spray the outer surface of the clothing. For your face, spray on your palms first, then apply, avoiding your eyes and mouth. Use sparingly around your ears.

Reapply according to the product label (every 2–8 hours). Over-application doesn't increase effectiveness. Wash your hands after applying. Remove the repellent with soap and water at the end of the day.

Apply to children's skin using your own hands. Never spray directly on children, and avoid their hands, cuts, and irritated skin. Keep bug spray away from their mouths and eyes, and don't use DEET products on infants.

Applying Permethrin

Treat clothing at least 24 hours before use to allow the solution to bond with fabric fibers. Apply to outdoor gear: clothing, shoes, socks, tents, backpacks, and sleeping bags. Keep wet permethrin-treated items away from cats. Use in a well-ventilated area. Let items dry completely before use.

Mosquito Prevention

Eliminate Standing Water to Reduce Mosquito Breeding Sites

Mosquitoes breed in standing water, so regularly inspect your property and eliminate potential breeding sites by:

- Scrubbing bird baths weekly to remove mosquito eggs and changing the water regularly.

- Turning over, covering, or disposing of items that can hold water (tires, buckets, cans, flower pots).

- Cleaning out gutters to prevent leaves and debris from blocking water flow.

- Filling tree holes with dirt or sand to stop water accumulation.

- Changing pet water bowls daily, especially during mosquito season.

- Maintaining swimming pools with proper chlorination and filtration.

Create Mosquito-Unfriendly Environments

- Install or repair window and door screens to keep mosquitoes outside.

- Use oscillating fans on porches and patios; mosquitoes are weak flyers and struggle against moving air.

- Plant mosquito-repelling vegetation around outdoor living areas like citronella, lavender, marigolds, and basil.

- Treat larger water features with mosquito dunks or larvicides designed for mosquito control.

- Keep vegetation trimmed to reduce resting areas for adult mosquitoes.

Tick Prevention

Yard Maintenance for Tick Control

Ticks thrive in tall grass, leaf litter, and brush. Good yard maintenance can significantly reduce their presence and limit your exposure.

Mow regularly. Keep grass shorter than three inches to eliminate tick habitats.

Remove leaf litter. Clear leaves, brush, and other debris from transition zones between wooded areas and lawns.

Create physical barriers. Use gravel or wood chip borders (at least three feet wide) to separate wooded spaces from frequently used areas, limiting tick migration.

Control host animals. Install deer fencing or plant deer-resistant vegetation to reduce the presence of deer that carry ticks into your yard.

Landscape Modification

Create dry, sunny areas around your home by pruning trees and shrubs to increase sunlight exposure. Remove Japanese barberry and other invasive plants that create favorable habitats for ticks. Some tick-resistant plants include:

- American beautyberry (Callicarpa americana)
- Sweet fern (Comptonia peregrina)
- Fragrant sumac (Rhus aromatica)

Consider installing cedar mulch in garden beds, which has natural tick-repellent properties, and place playground equipment, decks, and patios away from yard edges and trees.

CHAPTER 7
Food Safety

"Even the cleanest kitchen can't outpace a warming world." —Climate Cat

CLIMATE CHANGE AND FOODBORNE ILLNESS

You don't expect to get food poisoning from grandma's potato salad.

Yet here we are in a world where climate has turned our food safety assumptions upside down. Like an uninvited guest, foodborne illnesses appear in unexpected places, forcing us to rethink how we handle, store, and prepare our food.

Today, the FDA maintains critical food safety protections through its Human Foods Program, which was reorganized in October 2024. The agency currently conducts facility inspections, enforces safety standards, coordinates outbreak responses, and maintains traceability systems that help track contaminated foods from farm to table.

Under the Food Safety Modernization Act (FSMA), the FDA has shifted focus from responding to foodborne illness to preventing it, requiring food facilities to implement preventive controls and maintain detailed records.

In February 2025, many employees were terminated from the FDA's food division, including those working on chemical safety. Highly specialized staff who review potentially unsafe additives were among the layoffs. The potential disruptions to FDA's already strained inspection force are so great that agency leaders recently expedited plans to hire outside contractors to replace some fired workers.

These regulatory uncertainties become particularly dangerous in our warming world. Climate change is already increasing the frequency of extreme heat events, flooding, and power outages, which can compromise food safety. Higher ambient temperatures help bacteria like Salmonella and E. coli thrive. Floods can wash contaminants into crops or food processing plants. Power outages knock out refrigeration for hours or days, causing spoilage that isn't always visible or detectable by smell.

If comprehensive food safety oversight were dismantled, the consequences would compound these climate-driven risks. Without systematic facility inspections, unsanitary practices or spoiled products could reach store shelves undetected. Without coordinated outbreak response systems, dangerous foods could remain on the market for weeks or months before anyone sounds the alarm.

When multiple people in different states fall ill from the same source, it takes coordinated effort between the FDA, CDC, and local health departments to trace the cause, warn the public, and remove dangerous items from circulation. In a scenario where regulatory safeguards collapse, communication breakdowns between agencies and the public could delay life-saving information and worsen outcomes, especially for vulnerable populations.

Families would be left to navigate an increasingly risky and unpredictable food landscape without updated safety guidelines or emergency recalls. Climate change is making our food system more vulnerable to contamination just as political forces threaten to weaken the very systems designed to protect us. The simultaneous gutting of inspection staff and talk of dismantling agencies creates dangerous gaps in oversight.

In this environment, grandma's potato salad isn't the only thing you'll need to second-guess.

Each year, 48 million Americans learn that their food contains an unwelcome virus, bacteria, or parasite. That's 1 in 6 people sitting out some portion of their year with everything from mild discomfort to serious illness. While 128,000 of these folks end up in hospitals, around 3,000 don't make it home.[1] According to the World Health Organization, globally, foodborne diseases affect 1 in 10 people, resulting in 420,000 deaths annually.[2]

CLIMATE CHANGE
IS ALREADY
INCREASING THE
FREQUENCY OF
EXTREME HEAT
EVENTS, FLOODING,
AND POWER
OUTAGES, WHICH
CAN COMPROMISE
FOOD SAFETY.

Foodborne illnesses can be especially dangerous for those most at risk. As Ellen Ioanes and Li Zhou explain in *Vox*, "Listeria, salmonella, and E. coli infections are particularly concerning for the disproportionate effect they can have on vulnerable groups like pregnant people, children, and people over the age of 65."[3]

When Weather and Food Safety Collide

Climate change isn't just affecting our weather; it's also creating opportunities for foodborne pathogens. Most foodborne bacteria grow fastest between 40°F and 140°F, known as the "danger zone." For every 10°F increase in temperature, bacterial growth rates can double. Rising temperatures turn ordinary food storage into a breeding ground for microbes. Warmer springs and milder autumns have lengthened the warm-weather window in much of North America, creating more opportunities for pathogens to multiply.

Power outages during heat waves transform refrigerators into breeding grounds. Warmer conditions help bacteria survive for longer periods, and encourage toxin production in stored and canned items. Extreme weather events and flooding are like giving E. coli a free ride through our soil and water systems. In coastal areas, rising sea temperatures have increased Vibrio bacteria, making seafood safety increasingly tricky.

FOODBORNE ILLNESS	DESCRIPTION	IMPACT	PREVENTION
Norovirus	A highly contagious virus that causes vomiting and diarrhea. It spreads easily through contaminated food, water, surfaces, or contact with an infected person.	The annual impact includes 900 deaths (mostly among adults aged 65 and older), 109,000 hospitalizations, 465,000 emergency department visits (mainly young children), and 2,270,000 outpatient clinic visits.	Wash your hands thoroughly with soap and water, disinfect surfaces, and avoid preparing food when you are sick.

Salmonella	A group of bacteria often found in raw or undercooked meat, eggs, or unwashed produce. It causes diarrhea, fever, and stomach cramps.	Causes 1.35 million illnesses, 26,500 hospitalizations, and 420 deaths annually. It's the leading cause of hospitalizations and deaths from food.	Cook poultry, eggs, and meat thoroughly, wash your hands after handling raw meats or eggs.
E. coli	A type of bacteria that can live in the intestines of animals and humans. Some strains, like O157, can cause severe stomach cramps, bloody diarrhea, and kidney problems.	The O157 strain causes 96,534 illnesses, 3,268 hospitalizations, and 31 deaths annually. About 5–10 percent of cases develop hemolytic uremic syndrome, a serious kidney condition.	Cook ground beef to a minimum internal temperature of 160°F. Wash produce thoroughly, avoid unpasteurized milk and juices, and wash your hands after contact with animals.
Clostridium perfringens	An anaerobic bacterium that forms heat-resistant spores. Typically contaminates improperly cooked or stored meat, poultry, and gravies.	Causes a million illnesses, 438 hospitalizations, and 26 deaths annually.	Maintain hot foods at a temperature above 140°F and cold foods at a temperature below 40°F. Refrigerate leftovers quickly and reheat to at least 165°F.
Campylobacter	A bacterium commonly found in raw and undercooked poultry, causing diarrhea, fever, and stomach cramps.	Leads to approximately 1.5 million illnesses, 19,500 hospitalizations, and 128 deaths annually.	Cook poultry to at least 165°F, use separate cutting boards, avoid unpasteurized milk, and wash hands after handling pets.
Staphylococcus aureus	A bacterium producing heat-resistant toxins that cause rapid-onset vomiting.	Accounts for 448,400 illnesses, 1,451 hospitalizations, and 6 deaths each year in the U.S.	Wash hands before and during food preparation, keep foods below 40°F or above 140°F, avoid preparing food when sick.
Toxoplasma gondii	A parasite found in undercooked meat, posing severe risks for pregnant women.	Causes 86,686 illnesses, 4,428 hospitalizations, and 327 deaths annually.	Thoroughly cook meat, wear gloves when gardening, avoid raw meat during pregnancy, and clean cat litter boxes daily.
Listeria monocytogenes	A bacterium notable for its high fatality rate among foodborne pathogens, especially dangerous during pregnancy.	Leads to approximately 1,600 illnesses, 1,500 hospitalizations, and 260 deaths annually.	Avoid unpasteurized dairy, heat deli meats until steaming, keep the refrigerator below 40°F, and clean fridge surfaces regularly.

Like any crisis, foodborne illnesses disproportionately affect those who can least afford the impact. Pregnant women are ten times more likely to contract Listeria infections, and a simple food choice can result in miscarriage, stillbirth, or premature delivery.[4]

Young children, with their still-developing immune systems, are especially vulnerable, while seniors have higher mortality rates from norovirus because their immune responses are weaker.

People with compromised immune systems are at greater risk. HIV and AIDS patients face a twentyfold increase in serious illness, and for cancer patients or transplant recipients, each meal poses a calculated risk.

Simple Habits for Safer Meals

Hand Hygiene

- Wash your hands for at least 20 seconds with soap and warm water.
- Always wash before and after handling food—especially raw meat, poultry, eggs, or seafood.
- Use paper towels to dry hands to prevent cross-contamination.
- When soap and water aren't available, keep an alcohol-based hand sanitizer (at least 60 percent alcohol) nearby as a backup.

Setup and Preparation

- You can use a color-coded cutting board system to prevent cross-contamination:
  - Red for raw meat and poultry
  - Blue for seafood
  - Green for fruits and vegetables
- Clean surfaces, utensils, and cutting boards with hot, soapy water after each use.

Temperature Control

- Keep your refrigerator at 40°F or below[5]
- Keep your freezer at 0°F or below
- Cook ground meats and poultry to a minimum internal temperature of 165°F
- Cook fish and whole cuts of beef, pork, and lamb to a minimum internal temperature of 145°F, with a 3-minute rest time

If you suspect food poisoning, document everything you ate in the last 48 hours and note when your symptoms began. Keep any suspected food items and take photos of packaging or labels as evidence. Report the incident by calling your local health department and filing a report with the FDA at 1-888-SAFEFOOD (1-888-723-3366), or submit to the USDA at 1-888-MPHotline (1-888-674-6854) for meat or poultry concerns.

Seek medical care immediately if you experience bloody diarrhea, a high fever over 101.5°F, excessive vomiting, signs of dehydration, or symptoms that persist for more than three days. Proper documentation and reporting can help prevent others from becoming ill and assist healthcare providers in determining appropriate treatment.

"Wash Your Paws"

STAYING INFORMED ABOUT FOOD RECALLS

Staying informed about food recalls is crucial for protecting yourself and your family. Below are key resources for tracking food recalls:

- **FDA Recalls** • https://www.fda.gov/safety/recalls-market-withdrawals-safety-alerts
  Database of FDA-regulated product recalls

- **USDA Recalls** • https://www.fsis.usda.gov/recalls
  Information specific to meat, poultry, and egg product recalls

- **FoodSafety.gov** • https://www.foodsafety.gov/recalls-and-outbreaks
  Centralized resources covering all food types and recall information

To receive timely notifications about food recalls, consider these services:

- **Food Recalls and Alerts App**
    - https://apps.apple.com/us/app/food-recalls-alerts/id1508680654
    - https://play.google.com/store/apps/details?id=com.usfoodrecalls.android
  Resource for recall information

- **FDA Safety Alerts** • https://www.fda.gov/about-fda/contact-fda/get-email-updates
  Email notifications about FDA recalls and safety alerts

- **USDA Recall Updates** • https://public.govdelivery.com/accounts/USFSIS/subscriber/new
  Subscription service for USDA food recall information

- **Consumer Product Safety Commission** • https://www.cpsc.gov/Newsroom/Subscribe
  Alerts on product recalls, including food-related items

For comprehensive coverage of food safety issues:

- **Consumer Reports 10 Really Risky Foods Right Now** • https://www.consumerreports.org/health/food-safety/really-risky-foods-right-now-a7840705850/
  Source for food safety alerts and investigative reporting

- **Food Safety News** • https://www.foodsafetynews.com/

  Daily updates on food safety issues and recalls

- **Stop Foodborne Illness** • https://stopfoodborneillness.org/

  Nonprofit providing advocacy, education, and support for those affected by foodborne illness.

CHAPTER 8

Mental Health

"Whenever we bring the desire for the world's healing out into the open, whether through our actions or the groups we are part of, we help others do this too."[1]

WHERE IS MY REFUGE?

There are days when the air feels heavier than usual. Not from smoke or humidity. From something harder to name. An emotional gravity that drapes itself over our shoulders.

Climate change arrives in floods and firestorms. It also manifests as dread. A child sees a video of a wildfire and asks if their house will burn down. A teenager scrolls through collapsing ice shelves and quietly deletes their future college plans. A mother can't sleep during a heat wave, her baby sweating beside her, as the power grid flickers. A doctor breaks down in their car after another shift filled with breathing problems, panic attacks, and people who don't have anywhere else to go.

The climate crisis is a mental health crisis.

Children lack the language to express their grief. But they feel it. They know when the adults are worried. They know when something is wrong. Eco-anxiety in young people is rising with the temperature. According to a 2021 global study published in *The Lancet*, 59 percent of young people reported being very or extremely worried about climate change, and over 45 percent said it had a negative impact on their daily life and functioning.

Some kids act out. Some shut down. Others carry it like a quiet storm behind their ribs.

THE CLIMATE
CRISIS IS
A MENTAL
HEALTH CRISIS.

We need to give them space to speak and the tools to name what they're feeling. Climate change brings a sense of betrayal. We were told there would be systems to catch us. That someone somewhere was in charge. But then the floodwaters rise, the power goes out, and the ER has no backup air conditioning.

Here's something we don't hear enough: it's OK to grieve the world we're losing. It's OK to feel scared. Or angry. Or exhausted. You're responding, like any healthy person would, to a world in danger.

Mental health isn't a luxury. It's part of survival. Climate distress is real. When you live through floods, wildfires, and storms, when your home is washed away or burnt to the ground, when your life turns upside down, you will be changed. These calamities are layered on top of existing issues of anxiety, depression, loneliness, addiction, or thoughts of suicide. Climate change makes everything worse.

A word coined to describe the way people are affected by climate is *solastalgia*. "Nostalgia is the homesickness you have when you're away from home and wish to return. Solastalgia is the homesickness you have when you're at home, and your home is leaving you."[2] Imagine the house where you grew up, a park you love, or your favorite birds disappearing. Instead of simply missing them, you'd feel an ongoing sense of loss, of sorrow, because they're dying in front of you.

In this sea of change (environmental, political, and psychological), we need a refuge. The Buddhist concept of "refuge" refers to turning toward something we can rely on or rest in. That we can trust for comfort, guidance, and support during the most difficult times.

Buddhists rely on three key sources, known as the "Three Jewels": the Buddha (the teacher), the Dharma (the teachings or the path), and the Sangha (the community of fellow practitioners). We can do something similar for ourselves.

Taking Refuge in Life and the Earth: Recognizing Earth as our ultimate guide, understanding that our life, well-being, and survival depend entirely on the planet's health and ecosystems. Earth embodies wisdom, interconnectedness, and resilience, teaching us through its natural systems, balance, and immense beauty.

Taking Refuge in Grounded Actions and Strategies: Committing to sustainable, conscious practices and embracing clear strategies that help us heal ourselves and the environment. It's doing things, both personally and collectively, that restore balance, reduce harm, promote regeneration, and help us find peace.

Taking Refuge in the Global Community of Climate People and Activists: These are the advocates, scientists, climatologists, visionaries, artists, healthcare providers, and Indigenous communities around us. They remind us that we're not alone and that collective action works. What is *my* refuge?

The crisis is here. I've felt the pressure of fear, not knowing what's coming, and moments (or days, sometimes weeks) of hopelessness. The constant barrage of bad news feeds anxiety, despair, and even paralysis. I wasn't always an author. I wasn't always informed about climate change. And I certainly wasn't always calm.

But step by step, I learned.

My journey has been one of change, survival, and finding ways to keep going even when life feels impossible. I've discovered approaches that help me navigate my "inner climate" without shutting down. Everyone is different, and no single method always works. Having multiple tools means there's always something to try.

Here are the things I have personally found useful:

Yoga Nidra: A deeply calming practice done lying down, moving through a guided meditation focusing on body sensations like heat, cold, heaviness, and lightness. It induces a state of deep relaxation while keeping the mind aware. It's also very useful for sleep problems.

Meditation: I meditate every morning. One of my teachers said, "the best form of meditation is the one you do." I am drawn to meditations that give the mind a task, like those found in the practices of Tantra and

Taoism. A morning meditation helps set the tone for the day, preventing me from spiraling into anxiety. Insight Timer has many guided meditations. (https://insighttimer.com/)

Grounding and Being Outside: For me, being outdoors, especially by water, is one of the best ways to reset. I've always loved watching the transition of trees during the spring as they bud and grow leaves. Just working outside is a way of calming my nervous system.

EFT/Tapping: Tapping on acupressure points while repeating affirmations can help release anxiety. It's simple, accessible, and can be used in moments of high stress. The app I use is The Tapping Solution. (https://www.thetappingsolution.com/)

Exercise: Stress and anxiety are forms of energy that need to be released. Like meditation, the best form of exercise is the one you do. I like exercising in VR, Les Mills, walking in nature, and working out at the gym.

Music: Calming or uplifting music can help shift your emotional state. I listen to ambient music on Insight Timer. I also use Amazon Music, where I listen to heavy metal to get me energized, and classical music, particularly Franz Liszt, for focus.

Learning: Learning is a superb way to overcome depression or inertia. I've used Coursera, Udemy, and Mindvalley. Find what fascinates you.

Therapy: This may seem obvious, but experiencing climate change, fascism, and fear at the same time requires specialized support.

Gratitude: Kristi Nelson's *Wake Up Grateful: The Transformative Practice of Taking Nothing for Granted* shares the benefits of living with gratitude as a daily practice. Drawing on her experiences as a cancer survivor, she emphasizes the importance of embracing the present moment and appreciating life's simple joys, even in the face of our greatest fears.

She writes, "Gratefulness faces squarely that wonder and suffering live side by side, as do disaster and beauty. Grateful living helps us to never forget that we make the biggest difference through remaining connected to what we treasure, what we want to protect, and what truly matters most."[3]

The Evolutionary Force

Humanity has faced and overcome the devastations of wars and plagues, the fury of volcanoes and earthquakes, and the deluge of floods. In every era, in every evolutionary form—human, ape, primate, and back through all the myriad avatars of life—we have persevered. Our collective journey is one of tenacity, will, and adaptability.

When I feel defeated, I remember the legacy of Life, which is one of hope, learning, and strength. We survived every disaster. We will navigate this new one with the same spirit, cooperation, and creativity that carried us through the past. As star children and branches of the Tree of Life, we will rise to meet *this* moment.

We will protect our home, nurture our planet, and ensure a thriving future. Our hope and strength are in our unity and shared history. We will survive this, as we have survived everything else.

Engaging in Collective Action (Community)

The opposite of helplessness isn't "power." The opposite of helplessness is being helpful. Working with my publisher and editor on this book has been a collaborative effort. It's been more than a writing project; it's been a dialogue, reflection, and a course in learning how to communicate climate and health issues with clarity and compassion. I'm also in a community of authors, all working to share our messages with broader audiences and advocate for change.

As part of my research, I became involved with the CERT (Community Emergency Response Team) and the American Red Cross. These two organizations train you to respond to crises, while also creating a deep sense of community. Training sessions connected me to a network of people committed to helping others, reinforcing the idea that preparedness is not just an individual responsibility but a group one.

Another source of inspiration has been connecting with the Bluesky climate community. It's a world of poets, economists, scientists, and healthcare providers focused on the planet's future. According to Axios, "Action leads to hope. When it comes to facing climate anxiety, giving people agency can counterbalance stress and worry."[4] I've found this to be true in my own experience.

You don't have to be an expert to take action. Or a superhero to make a difference to others. What you can do is take care of yourself and your family, find your place in this movement, and move forward with clarity and purpose.

Project Drawdown is a comprehensive research initiative that maps 93 practical solutions for reducing greenhouse gases and reversing global warming. Climate action is accessible to everyone because there are entry points everywhere, whether you're interested in gardening, technology, policy, education, or community work. You don't need to be a climate scientist or policy expert to contribute meaningfully to climate solutions.

You can start with anything that's important to you: trees, the circular economy, plastics, education, health, politics, community gardening, energy, or food banks. There is always *something* you can do. (https://drawdown.org/solutions/table-of-solutions)

The world we're moving into requires us to be conscious, intentional, and resilient. Whether parenting, teaching, organizing, or being good neighbors, we can all care for one another. You have something unique to contribute; you are the solution. Take care of yourself, support your community, and step into the future with strength and purpose.

*Climate Cat's
Concern
For the
Unprotected*

CLIMATE CAT'S MENTAL HEALTH TIPS

Paws: Before you feel safe in the yard, you must know where the evil dogs are.

Collar the Wind: Let your care and concern shape the world, not just echo through an empty house. Give play to your intensity. Combine your collective power to nourish and heal the earth.

Tail Twitch Together: There's hope and a way forward in unity.

Chasing Tin Foil Balls: From fumbling in the dark to the awakening spark. From the shock of realization to the meow of outrage. And finally, a courageous leap into action.

Every Cat Is a Catalyst: Embrace your role as a catalyst: every action, every voice, every choice builds an entire future.

We Need Each Other

Hexagram 29 of the I Ching is K'an – The Abyss (Water). This hexagram consists of water doubled: danger upon danger, depth upon depth. It represents being in the middle of genuine crisis, yet finding a way through by maintaining inner clarity and staying true to your course.

- **Acknowledging real danger:** the crisis is here
- **Moving through difficulty:** it's unavoidable
- **Maintaining inner light:** even in the deluge
- **Water's teaching:** flow around obstacles, fill the low places, and keep moving forward
- **Sincerity and steadfastness:** in the face of the abyss

The world needs you.

Your presence, your clarity, your peculiar gifts. It needs you rested enough to think clearly. Grounded enough to act wisely. Connected enough to remember who you're fighting for. And why.

The crisis asks us to be wise, strategic, and creative. To build something that lasts longer than our initial surge of panic or despair.

Tend to yourself. Build your refuges. Find your people. Learn your practices. And when you're ready, step forward. The work is here. The community is welcoming. Together, we'll do what humans have always done when faced with the abyss.

We'll adapt. We'll create. We'll care for each other.

And we'll keep going.

COMMUNICATION AND MEDIA

CHAPTER 9
Signals and Sirens

"The strongest signal in a crisis is trust."

—Climate Cat

USES OF SOCIAL MEDIA DURING EXTREME WEATHER

Social media played a crucial role in communication and coordination during the devastation caused by Hurricanes Harvey, Irma, and Maria. After the 2017 hurricane season in the United States, social media "hosted millions of requests for information, communication between family, friends, and strangers, calls for donations, spaces to share collective grief and compassion, and rescue requests and pleas for immediate help."[1]

USAGov utilized social media channels to disseminate preparation and recovery information, with a focus on shelters, victim assistance, and safety measures. During the 2021 heat wave in the West, social media played a crucial role in communicating the locations of cooling centers and raising awareness about heat illness symptoms. Many agencies, counties, and cities shared and amplified messaging in multiple languages to provide information and resources to their communities.

Social media can serve as hubs for real-time updates during extreme weather events. By preparing in advance, you can quickly receive and share critical information during emergencies. You then become part of a global network of "on-the-ground" reporters, contributing to the flow of validated news. By sharing stories of recovery and positive action, you can spread hope during the crisis.

IN A DISASTER,
ATTENTION SPANS
ARE SHORT, AND
STRESS IS HIGH.
COMMUNICATE IN
SIMPLE, DIRECT
TERMS THAT ANYONE
CAN UNDERSTAND.

Preparing Your Social Media for Emergencies

Update Your Social Media Profile Contact Information

Indicate any emergency roles or details. For example, include a phone number or website in your bio so people know how to reach you or find more information during a crisis. Up-to-date profiles lend credibility and make it easier for others to get help and verify your identity.

Follow Key Emergency Organizations and Experts.

This includes local emergency management agencies, public health departments, weather services, and climate scientists. These sources provide timely alerts and accurate updates.

Rely on Official Websites and Verified Accounts

By having trusted sources in your feed, you can quickly share their posts to amplify crucial alerts (for example, retweet your city's evacuation notice or share your local health department's heat advisory).

Establish Credibility

If you represent an organization or community group, verify your accounts whenever possible. A verified status (or other proof of authenticity) reassures the public that your updates are official and trustworthy.

Double-Check Information from Others Before Sharing It

Pause and avoid sharing impulsive "gut reactions." In emergencies, rumors spread quickly. Take a moment to verify facts via trusted news or official channels. By sharing only confirmed information, you help prevent the spread of misinformation.

Use Clear Language

In a disaster, attention spans are short, and stress is high. Communicate in simple, direct terms that anyone can understand. Stick to the facts: who, what, where, when, and how. Avoid commentary.

If you need to share a detailed official statement, post a summary or key point and provide a link to the full text. The goal is to inform, not overwhelm. Clear and concise messaging keeps people informed without causing confusion.

Cite and Share Official Info

Whenever possible, back up your posts with information from official sources (such as linking to a government webpage or including a graphic from an authoritative agency). For example, you might share a FEMA infographic on emergency kit supplies or a National Weather Service graphic on heat safety. Attaching official graphics or links makes your post more informative and signals that the content is vetted.

Many agencies provide ready-to-share content for this reason. The National Weather Service even invites the public to share their safety messages and images to spread the word. By amplifying these resources, you help circulate accurate guidance.

Post Regularly During Crises

When an emergency occurs, commit to providing frequent updates if you're in a position to offer ongoing information (such as an administrator of a community page or a local official). Let your followers know you will be providing updates and at what interval.

Consistent communication, even if just to say "no change, stay tuned," reassures people that someone is paying attention. Officials suggest that actively publishing frequent updates promotes transparency and helps "control the message" before rumors fill the void. If you can't keep up that pace, update on a regular schedule (every morning and evening). And if you learn something important between scheduled posts, share it as soon as possible.

What to Share

Effective social media use during a disaster involves knowing *what* to share and *how* to present it to maximize reach and comprehension.

WHAT TO SHARE	HOW TO SHARE
Current health risks (heat index, air quality, flood conditions)	Text updates with emojis and hashtags (Heat Advisory: Heat index 105°F. #HeatSafety), screenshots, or links to live data from official sources.
Symptoms requiring immediate medical attention	Infographics comparing symptoms (Heat Exhaustion vs Heat Stroke), and short videos with first aid instructions.
Locations of emergency medical services	Interactive Google Maps with pinned locations, pinned posts on Facebook or X listing addresses and contact numbers.
Mental health support resources	Carousel posts listing hotlines and resources, as well as Instagram stories with swipe-up links to counseling services.
Available Telehealth options	Step-by-step guides in threaded tweets or multi-image posts, live Q&A sessions explaining access to virtual care.
Maps of cooling centers, air quality, medical wait times, evacuation routes	Custom or embedded maps shared with clear legends, links to interactive versions.
Infographics about heat illness, air protection, disease prevention, preparedness tips	Platform-optimized images with alt text, reposts from trusted sources with credit.
Climate-resilient shelters with medical support	Cloud-based documents with real-time updates, shared links and pinned threads.
Food and water distribution centers	List posts updated daily, infographics showing hours and supplies available.
Healthcare locations offering climate-related care	Highlight stories on Instagram titled "Medical Resources," reposts from health departments.
Mental health hotlines for climate anxiety	List of phone/text hotlines with icons, and carousel graphics for emergency coping strategies.
Stress management tools for extreme weather	Daily posts with coping tips, video reels on breathing exercises or mindfulness.

Platform-Specific Strategies

Platforms change fast, including ownership, algorithms, and audience behavior. Misinformation spreads like fire. It's essential to use social media with full awareness of who owns it and how it shapes everything we see.

Different audiences gather in different spaces. Each platform serves a purpose, and that matters during a crisis. Below is a simple guide on using popular platforms effectively during a climate emergency.

X (Twitter)

Useful for fast, real-time updates. Follow official sources and verified accounts. Share weather alerts, air quality warnings, symptoms to watch for, and shelter locations.

Use hashtags to make your post easier to find, like:

- #HeatSafety, #AirQuality, #FloodAlert
- Event-specific tags like #HeatWave2026
- Location tags like #NYCFlood or #LAHeat
- Don't overdo it: Stick to two to three clear hashtags per post.

Instagram

Good for visual content and quick learning. You can share:

- An infographic showing the air quality index scale with color codes, explaining what "Very Unhealthy" air means for outdoor activities.
- A chart of heat index values and corresponding health risk levels. A single image comparing "Signs of Heat Exhaustion vs Heat Stroke" with icons and bullet points can quickly educate viewers.
- Images of maps that mark important facilities like cooling centers, emergency shelters, medical clinics, and supply distribution points. During an extreme heat event, a map of the city with pins on all the cooling centers helps people instantly visualize the nearest location.

Maps make it easier for followers to plan where to go if they need relief or assistance. Ensure these maps are updated (if an official map is available from your city or Red Cross, share that for reliability).

- Use carousel posts to show step-by-step guides or multiple resources. Use alt text to make images accessible.

- Many health agencies offer pre-made, shareable graphics. Reposting those can ensure accuracy. Some examples from the California Department of Public Health: https://www.cdph.ca.gov/Programs/OPA/Pages/Communications-Toolkits/Heat-Summer-Safety.aspx

Telegram and Signal

Local volunteers might form a Telegram group to arrange food and water distribution, share shelter locations, or coordinate transportation for evacuees. One recent example comes from a California wildfire: Telegram groups emerged where residents coordinated volunteer efforts, shared resources like supplies and generators, and provided real-time eyewitness reports when official information was scarce.

WhatsApp

Form neighborhood check-in groups. Conduct wellness checks on seniors and vulnerable neighbors. Share who has power, water, or a working phone. Coordinate visits and supplies.

A simple message like "Has anyone checked on Mr. Lee in Unit 5?" can be life-saving during heat emergencies or winter storms when isolated individuals might be at high risk.

Facebook Groups

These are useful for quickly building community support. After Hurricane Harvey, one local group grew to 130,000 members and became a vital lifeline for people stranded or in need of supplies.

You can:

- Join existing emergency groups.
- Start your own for your town or apartment complex.
- Share official updates, post resource links, and offer help.

Reddit

Engage with relevant communities and join subreddits specific to your region or those focused on climate issues, such as r/climate and r/preppers. Participate in discussions about climate preparedness and response. Share and source information:

- Post accurate information on local climate conditions, emergency procedures, and resources. Ensure the credibility of your sources to maintain trust within the community.
- Gather real-time information on road closures, air quality, and local support services.
- Coordinate assistance efforts by connecting with others who are offering or seeking help.
- Share guides, checklists, and personal experiences related to climate preparedness.

Discord

- Can be used as a supplementary tool for tech-savvy audiences.
- Establish or join climate-focused servers.
- Platforms like DISBOARD list various climate-related servers, such as the "Discord Eco-Network," which connects multiple environmental communities. (https://disboard.org/servers/tag/climate-change)
- If a specific need arises, consider setting up a new server focused on particular climate events or local initiatives.
- Structure the server with channels tailored to different topics, such as:

- #emergency-alerts: For real-time updates on climate events.

- #resource-sharing: To post and find supplies or services.

- #volunteer-coordination: To organize community response efforts.

Social media can be life-saving, but it's imperfect. Algorithms can bury critical info. Disinformation can mislead. And access isn't equal: some people still rely on phone calls, radio, or word of mouth. Use these tools wisely, check your sources, and think locally. Know your platforms. Know your neighbors. And most importantly, stay connected.

CLIMATE CAT'S GUIDE TO MANAGING
INFORMATION OVERLOAD

The media bombards us with images of flooding, destruction from hurricanes, and orange skies from forest fires. Since so much of climate news is "bad news," we need a plan for media consumption:

1. Schedule specific times to read or watch the news rather than checking continuously throughout the day.

   Limit Screen Time: Set a daily time limit for social media use. Studies show that reducing social media use to 30 minutes a day can decrease anxiety, depression, and loneliness.

   Schedule Breaks: Take regular breaks from social media, especially first thing in the morning and before bed.

   Turn Off Notifications: Disable social media notifications to reduce the temptation to constantly check your phone.

   Time-Box: Check social media at designated times rather than throughout the day, to help limit compulsive checking.

   Use Alternative Devices: Access social media only from your computer, not your phone, to make it less convenient to use impulsively.

   Set Boundaries: Clearly define and communicate your social media boundaries to friends and family to maintain balance.

2. Be mindful of your emotional responses, and practice self-care techniques to manage stress and anxiety. When you feel overwhelmed, limit your media consumption and engage in activities that promote well-being, such as meditation, exercise, or spending time in nature.

Tech-Free Time in Real Reality: Implement tech-free weekends or certain hours during the day to reconnect with real-life activities and people.

3. Follow reliable sources that provide balanced coverage, including solutions and positive action. For example, climate scientist Katherine Hayhoe shares both good news and not-so-good news in her weekly newsletter: https://substack.com/@katharinehayhoe.

4. Curate feeds. Unfollow accounts that make you feel inadequate, hopeless, or endlessly enraged. Not all rage is bad. If it fuels action, builds movements, or motivates change, it can be powerful. But if it leaves you stuck, spiraling, or frozen, it's just another meme-wave crashing into your nervous system.

Even Climate Cat Needs to Take a Break

5. Seek out the voices that energize you, inform you, and remind you of your agency. Follow those who inspire and uplift, who offer solutions alongside problems.

6. Tools like Flipboard or Feedly can help customize your newsfeed to highlight stories of progress, innovation, and resilience: so you're not just surviving the scroll, but shaping it.

7. Follow organizations and individuals dedicated to climate solutions, innovations, and community engagement. Examples include Project Drawdown (https://drawdown.org/) and the Climate Foundation (https://www.climatefoundation.org/).

8. Diversify your sources to include those that offer in-depth analysis, scientific perspectives, and stories of hope and resilience. Websites like Inside Climate News offer investigative reporting, climate science, and stories about climate action.

9. Join online or local groups focused on climate activism and solutions. Engaging with communities working toward positive change can provide a more balanced perspective.

INFORMATION AND DISINFORMATION

"We are all responsible for what we hold to be accurate and for defending that truth against attack."[1]

The words *misinformation* and *disinformation* are often used interchangeably.

Misinformation is false or inaccurate information. Disinformation is intended to mislead and confuse, i.e.,propaganda. Even though these words have two distinct meanings, a lie is a lie. Extreme heat, pollution, wildfires, ticks, and mosquitoes are deadly. So is health disinformation.

The internet allows unfiltered access to everything: reliable sources, information warfare, and lies. Just as there are physical storms, there are storms of human stupidity. The United States currently has the second-highest rate of misinformation in the world.[2]

Politicians, websites, and books deny that fossil fuels cause climate change, attack science and scientists, and spread conspiracy theories about environmental policies. At the time of this writing, climate science, food safety, public health, and meteorology are all under attack. Politicians are banning discussions about climate change and trying to eliminate the National Oceanic and Atmospheric Administration (NOAA).

But the attacks go far beyond the environment.

Public health and medicine have become prime targets in this new era of weaponized disinformation. Anti-vaccine rhetoric, once on the fringe, has moved into the mainstream, amplified by social media and elected (and non-elected) officials. Life-saving vaccines and basic medical practices are now treated as partisan issues. Diseases that were eradicated are making a comeback. Measles is no longer a historical footnote, it's back in the headlines.

The Centers for Disease Control and Prevention (CDC) and National Institutes of Health (NIH), long respected globally, are being undermined from within and without. Their budgets are slashed, their authority questioned, and their recommendations politicized. Doctors and researchers face harassment and threats for doing their jobs. Science is rebranded as ideology, and evidence as opinion. Medical professionals are shouted down by conspiracy theorists with YouTube channels.

This isn't just a war on facts. It's a dismantling of the institutions designed to protect life and health.

Agencies that monitor disease outbreaks, ensure clean air and water, and support emergency response are being hollowed out. Expertise is being replaced by ideology. The result? A nation less prepared for pandemics, unable to respond to natural disasters, and more vulnerable to both nature and human error.

Disinformation is a public health crisis, an environmental threat, and a national security issue. Lies, left unchecked, harm and kill.

The links between human caused climate change, politics, and misinformation are not accidental, they are part of a coordinated strategy. A powerful alliance of fossil fuel interests, political actors, and ideological media works to discredit science, dismantle regulatory institutions, and sow public confusion. By attacking the EPA, HHS, Medicaid, climate scientists, and healthcare professionals, they aim to erode public trust in facts, weaken science-based policy, and protect profits and power at the expense of public health and planetary stability.

These efforts continue as the severity of climate events (and their health consequences) intensifies. Extreme weather, pollution, and the spread of disease are on the rise, but so is the onslaught of denial and deception. The result is a vicious cycle: as climate impacts worsen, the spread of health disinformation accelerates, making it ever harder to protect communities from the very threats these interests downplay.

Just as drug dealers need a distribution network, social media is a global distribution network for misinformation. Most Americans now rely on social media for at least part of their news. Misinformation is like wildfire; it spreads "significantly farther, faster, deeper, and more broadly than the truth in all categories of information."[3]

DISINFORMATION IS A PUBLIC HEALTH CRISIS, AN ENVIRONMENTAL THREAT, AND A NATIONAL SECURITY ISSUE

Disinformation Causes Death

A recent example is the 300,000 preventable COVID-19 deaths in the United States. Nearly one-third of all COVID deaths were people who would be alive if they'd gotten vaccinated. Without reliable information or governance, disinformation will convince millions to reject life-saving interventions and embrace harmful practices.

"Only 14% of the United States population has proficient health literacy, which makes it difficult to recognize medical misinformation."[4] In a climate-stressed world, people are misled into avoiding vaccines, taking ineffective or dangerous treatments, or rejecting essential public health measures. Epidemics that could have been contained will spiral out of control, fueled by disinformation. "Health misinformation could delay proper treatment, which could further deteriorate patients' health status and affect relevant outcomes, including mortality rate."[5]

The same forces—people, organizations, and industries—that created the climate crisis are not going to let us solve it. As weather extremes intensify, so will the flood of lies, propaganda, and manipulation.

During Hurricane Helene in the fall of 2024, people avoided help from FEMA because they were told it was a scam. Fossil fuel interests perfected the playbook for manufacturing uncertainty and mistrust. These same tactics are increasingly applied to health disinformation, whether to sow doubt about climate-induced diseases or to undermine public trust in sustainable technologies.

By obscuring the health consequences of pollution, extreme heat, and environmental collapse, fossil fuel operatives deflect blame and stall action. Their influence goes far beyond climate denial; it seeps into the broader public discourse, fueling a deep mistrust of science, expertise, and the institutions meant to protect us.

"Truth Isn't Always Loud, Sometimes It Needs to Be Found"

CLIMATE CAT'S SEARCH FOR THE TRUTH

⚡ Since facts (truth) and disinformation (lies) are so deeply interwoven, how do we distinguish them? How can we protect our mental and physical health when the information we rely on is fragmented, manipulated, politicized, and distorted?

⚡ In an era where "data is the new oil," developing *media literacy* is necessary for our health. Media literacy helps us to analyze, evaluate, and understand media in all its forms. We can then interpret media messages critically, revealing the deeper processes and intentions.

How-To Guide for Media Literacy

Fighting misinformation doesn't mean you have to scrutinize every post or become a full-time fact-checker. Sometimes, the red flags are apparent: sensational headlines, emotional manipulation, or wild claims with no evidence. Other times, the lies are more subtle. The key is to stay curious, not cynical. Sources, scientists, and journalists are working hard to get it right.

Start by asking what the message wants from you: your outrage, your money, or your loyalty. Who benefits if you believe it? Notice what's missing, whose voices are absent, and whether the story fits into a larger political or economic pattern. Some narratives aren't just careless: they're coordinated.

The credibility of a message often depends on its source. Use platforms like *AllSides* to understand how political bias shapes media coverage and cross-reference information with organizations like *ProPublica* or *The Associated Press*, known for their journalistic integrity.

When in doubt, verify claims across multiple outlets. Apps like Flipboard and Feedly help you track reliable news from diverse perspectives. For specific claims, especially in politics or science, use fact-checking platforms: FactCheck.org and Snopes are leaders in investigating misinformation. Check dates carefully; outdated information can be just as misleading as falsehoods.

Health and science evolve quickly, and recycled headlines create unnecessary panic. For up-to-date, peer-reviewed health data, visit PubMed or MedlinePlus. Vet the author. A quick look at someone's LinkedIn or Google Scholar profile can tell whether they're an expert or just playing one online.

Check for conflicts of interest. If someone has ties to fossil fuel companies, political think tanks, or advocacy groups, those affiliations matter. Desmog's Climate Disinformation Database lists the backgrounds of individuals spreading disinformation.

Some disinformation can be visually manipulated. Be skeptical of memes and "screenshots" that spread without context. Use tools like FotoForensics to analyze images.

If something feels suspiciously well-timed, popping up during elections, major climate disasters, or legislation debates, it probably is. Who's funding the content matters. OpenSecrets.org shows corporate political donations, while LobbyView helps track who's paying to push policies. Use ProPublica's Nonprofit Explorer to check whether industry interests quietly fund a supposedly independent source.

There's no vaccination against disinformation, but there are ways to strengthen our immune systems. *Climate Central*, *EcoWatch*, and *The Conversation* provide honest, science-based reporting on climate and environmental issues. For health, the CDC, NIH, and WebMD remain references.

Archived versions of government web pages can be accessed through several platforms that preserve official content for historical and research purposes. These include https://govarchive.us/, which captures and stores snapshots of U.S. government websites; https://archive.org/, home of the Internet Archive's Wayback Machine, which provides extensive archival coverage of public digital content; and https://restoredcdc.org/www.cdc.gov/, which offers a preserved version of the CDC's website.

If you want to build stronger habits around media literacy, the News Literacy Project offers practical training for the digital age. Critical thinking is a survival skill. Knowing how to evaluate information protects our communities, our health, and our democracy.

NAVIGATING HEALTHCARE

CHAPTER 10

Health Insurance 101 for Climate Emergencies

"Coverage you understand becomes care you can reach." —Climate Cat

HEALTH INSURANCE LITERACY

The middle of a hurricane, power outage, or blizzard is not the time to be fumbling to find your health insurer's website password, or realizing you never registered and don't know how to use their site. Being prepared in advance for your healthcare makes all the difference during a weather emergency.

Let's think of "health" as a series of skills we build over time.

First, health literacy refers to the ability to read, understand, and apply health information effectively.

Then there's *digital health literacy*, which involves using technology such as electronic health records, health apps, digital platforms, and wearable devices to access and manage your health information.

Climate health literacy is knowing how climate change affects you, identifying risks, and adapting care strategies.

Health insurance literacy is understanding, navigating, and managing your health insurance coverage. It includes:

- Using your insurer's online portal and mobile app
- Accessing and understanding the details of your policy

- Knowing which services are covered, how to review your benefits, and how to take advantage of programs your plan offers

- Finding in-network doctors, hospitals, and urgent care centers

- Estimating your out-of-pocket costs, such as deductibles, copays, and coinsurance

- Using cost-comparison tools, avoiding surprise bills, and resolving coverage issues

- Managing the mess of the claims process: tracking claim status, knowing what to do if a claim is denied, and how to file an appeal

Very few people are health insurance literate. Research indicates that many Americans struggle with fundamental concepts of health insurance. I've met people who work in health insurance who aren't aware of their coverage or costs.

Remember that *more extreme weather will lead to more health problems, which lead to more doctor visits and hospitalizations.* Extreme weather puts a huge strain on our healthcare system. Emergency rooms fill up with people suffering from heat stroke. Ambulances are called for severe dehydration, heart issues, or trauma. Many end up in the hospital with heat-related illnesses, serious respiratory problems, or infections from floodwaters.

Medications like inhalers, antibiotics, and insulin become even more essential, along with specialized care such as burn treatment, dialysis, or surgery. Mental health needs increase, particularly for individuals coping with the emotional stress of disasters. People with chronic conditions like diabetes or heart disease face even greater risks when their regular care is disrupted. Individuals undergoing treatment for opioid use often lose access to their buprenorphine medication.

Climate change will make all of these more frequent.

Knowing how to navigate your health insurance and the healthcare system *before a crisis* will make a difference in getting the care you need *during* one.

KNOWING HOW TO NAVIGATE YOUR HEALTH INSURANCE AND THE HEALTHCARE SYSTEM *BEFORE A CRISIS* WILL MAKE A DIFFERENCE IN GETTING THE CARE YOU NEED *DURING* ONE.

General Preparations

Register Your Account: Use your insurance ID card to create an online account with your insurer. Test your login credentials to ensure access during emergencies.

Download the Mobile App: Install your insurer's mobile app on your device and keep it updated for quick access to information and notifications.

Explore the Website and App Features: Familiarize yourself with your insurer's website layout and app features to easily find critical information and programs during a crisis. Some insurers' portals are not user-friendly, even for experienced users. This is a good reason to test all features in advance.

Download and Save Your Digital ID Card: Keep both physical and digital copies of your insurance card and contact details for emergencies.

Save the Customer Service Number: Store your insurer's customer service number in your phone for easy access.

Understand Your Coverage and Network: During an emergency, access to in-network care, medication refills, or urgent services may be limited. Knowing your options in advance can help prevent costly delays or denials.

Finding Care

✓ Identify nearby in-network doctors and facilities. Knowing your in-network options in non-emergency situations helps with follow-up care and ongoing costs. Start by finding the nearest in-network hospitals and urgent care centers.

✓ Use search tools. Your health insurance website or app will typically feature a "Find Care" or "Doctor and Hospital Finder" option to help you locate in-network doctors, urgent care facilities, and hospitals. Keywords like "emergency" or "urgent" can help. Important caveat: data can be incorrect; double-check that a provider is in your network by calling your insurer.

✓ Locate multiple nearby hospitals, urgent care centers, and pharmacies. This ensures options if your nearest facility is closed or overwhelmed, especially during emergencies or climate events.

✓ Save key contacts for urgent care and ambulance services.

✓ Set up telehealth accounts in advance to access virtual care during emergencies. You can use your insurer's telehealth services, your doctor's virtual care offerings, or platforms including Teladoc and Amazon Prime for telehealth support. Telehealth often covers general care, mental health, allergies, and dermatology.

✓ See if your insurer offers free or low-cost virtual mental health services. These provide therapy, coaching, and psychiatric care, and can help you access support when in-person visits aren't possible.

✓ Save your insurer's 24/7 nurse hotline number for non-critical guidance.

✓ Familiarize yourself with services offering second medical opinions for complex health issues.

Understanding Emergency Coverage

Insurance policies typically outline emergency coverage in specific sections titled "Emergency Services," "Emergency and Urgent Care," or "Catastrophic Coverage." Look for terms like "natural disasters" or "national emergencies." Understand what your plan includes regarding deductibles, copays, coinsurance, and out-of-pocket maximums.

A deductible is the amount you must pay before the insurer starts covering costs; in emergencies, this might be waived or reduced, depending on the insurer and the nature of the disaster declaration (federal vs. state).

Copays are fixed amounts for services like ER visits, and they may change or be waived during disasters.

Coinsurance is your share of costs after meeting the deductible, often shown as a percentage; verify whether this changes during emergencies.

Your out-of-pocket maximum is the most you will pay in a policy year for covered services. Check if emergency care during disasters counts toward this limit and whether there are claim caps.

Contact your insurer to clarify pre-approval requirements for emergency services. While non-emergency care usually requires pre-approval, these rules are often lifted during a crisis. Ask about any disaster-specific exceptions for events such as hurricanes or wildfires.

Review your policy for waivers on out-of-network care, which are often granted when access to in-network providers is limited. These provisions may appear under sections like "Emergency Medical Transport," "Hospitalization," or "Catastrophic Coverage." Some insurers include a dedicated disaster clause or rider that allows coverage for out-of-network providers during a declared emergency.

Confirm whether your insurance remains valid if you must relocate temporarily and ensure coverage extends across state or regional lines.

Use your insurer's cost estimator tools to anticipate potential expenses for emergency care and prepare accordingly. Many cost estimator tools are imprecise. Transparency in pricing is still an evolving mandate, and many tools give only estimates, not guarantees.

Doctors

Ask your doctor for an emergency summary of your health records, including allergies, chronic conditions, and current medications, and store this information securely for easy access in case you need to see an unfamiliar doctor. Discuss your specific health needs with your doctor to adjust your emergency care plan, and set up a secure backup communication method, such as a secure messaging app, in case phone lines are unavailable.

Have contingency plans in place if your regular doctor is unavailable. This includes identifying backup healthcare providers, having telehealth options, and ensuring you have their contact information. Prepare a list of specialists you may need during an emergency and make sure their contact details are current.

AFTER THE DISASTER: HEALTH INSURANCE CLAIMS

Climate change leads to more health impacts, more doctor visits and hospitalizations, and more insurance claims. More claims lead to more claim denials, unexpected medical bills, growing medical debt, and even medical bankruptcy. The politicization and fragmentation of healthcare make these problems worse.

Americans now have over $140 billion in medical debt in collections, more than any other type of debt, according to a 2021 analysis by the *Journal of the American Medical Association*.[1] As claims grow, so does the need for expertise in disputing denials.

But wait, it gets worse:

- Treatment for injuries sustained during floods, fires, hurricanes, or heat emergencies may be denied if not properly documented or linked to a medically necessary diagnosis.

- When medications are lost during evacuations or due to home damage, insurers may deny replacement prescriptions as "early refills."

- Claims for heat stroke, heat exhaustion, and related complications may be denied if they are not properly documented as emergency conditions.

The No Surprises Act of 2022

Accessing medical care during a disaster can be frightening, especially if your usual healthcare providers are unavailable. The No Surprises Act of 2022 is a federal law designed to protect consumers from unexpected medical bills, called "surprise billing." The No Surprises Act helps protect you when you might need emergency care.

CLIMATE CHANGE
LEADS TO MORE
HEALTH IMPACTS,
MORE DOCTOR
VISITS AND
HOSPITALIZATIONS,
AND MORE
INSURANCE
CLAIMS.

Protection During Emergencies

Climate emergencies, such as hurricanes, wildfires, or floods, can disrupt access to your regular healthcare providers. You may need to seek care from out-of-network providers or unfamiliar emergency facilities. When you're facing a medical emergency, the last thing you should worry about is navigating confusing insurance rules or unexpected bills. Thanks to the No Surprises Act, important protections are in place to shield you from surprise medical costs and ensure you can access care when you need it most. Here's what you need to know:

You Can Receive Care at *Any* Emergency Department: The No Surprises Act ensures that you will only be charged in-network rates even if you receive emergency care at an out-of-network facility. Out-of-network providers cannot bill you more than in-network cost-sharing amounts (deductibles, copays, coinsurance).

No Prior Authorization Needed: In an emergency, you will not have time to navigate insurance bureaucracy. The law prohibits health plans from requiring prior authorization for emergency care, meaning you can focus on getting the care you need without worrying about whether it's covered upfront.

Post-Stabilization Care: After receiving emergency care, you may still require additional medical attention, which is referred to as *post-stabilization care.* The law prevents out-of-network providers from billing you excessive charges for this care unless you are fully informed and consent to waive the protections.

In-Network Billing Guarantees: Even if you're treated by an out-of-network provider at an in-network facility, the law guarantees you won't be charged more than your in-network deductible, copay, or coinsurance.

Clear Billing Information: The No Surprises Act requires providers to give you clear notices explaining your protections and the process for raising concerns if a provider violates these protections.

Recent administrative actions have introduced uncertainty regarding the effective implementation of the No Surprises Act. Budget reductions have raised concerns about Centers for Medicare and Medicaid Services's (CMS) capacity to enforce the act's provisions effectively. The *Independent Dispute Resolution* (IDR) process, a cornerstone of the act designed to mediate payment disputes between providers and insurers, has encountered operational problems. In 2023, over 650,000 disputes were filed, leading to significant backlogs and delays.

Protecting Yourself from Overbilling

The late Marshall Allen, in his book *Never Pay the First Bill: And Other Ways to Fight the Health Care System and Win*,[2] offers strategies to protect against hospital overbilling. He recommends writing a clause into financial agreements before signing: "I consent to appropriate treatment and (including applicable insurance payments) to be responsible for reasonable charges up to two times the Medicare rate." This clause helps protect against excessive charges. Marshall Allen says, "Everything is negotiable in healthcare billing." *See Appendix for detailed instructions on how to use the 2X medicare clause.*

The Marshall Allen Project provides free AI assistance, video-based curriculum, books, and access to trained Patient Advocates: https://www.marshallallenproject.org/. The Marshall Allen clone provides AI assistance on how to analyze medical bills for errors and unfair pricing, research reasonable costs for medical services, and contest inaccurate charges.

Appealing a Denied Claim

Most people simply accept claim denials. Consumers appeal fewer than 1 percent of claim denials, but statistics show the average success rate for appeals ranges between 50 percent and 60 percent, depending on your state and specific insurer. That 50 to 60 percent success rate for appeals means the odds are in your favor if you take action.

Knowing your rights and understanding the appeals process can help avoid financial shocks after extreme weather events. Stay calm, be prepared, and expect obstacles and delays.

Preparation

Document Everything: Maintain detailed records of dates, doctor names, reasons for visits or treatments, and any prescriptions. Track your expenses (copays, deductibles, out-of-pocket payments, and coinsurance). Record phone calls and note dates when speaking to insurance representatives. (Check your state's laws on recording conversations. Some require notifying all parties.) All of this documentation will help in disputing charges, filing claims, or ensuring continuity of care.

Access Your Medical Records: Many healthcare providers and hospital systems offer online portals, such as MyChart, where you can view your medical information, test results, medications, and appointments. Under HIPAA, you can inspect, review, and receive copies of your medical records. Providers must share your health information within 30 days of your request.[3]

Understand Why the Claim Was Denied: When you receive a claim denial, first check the explanation of benefits (EOB) that accompanies it. Every EOB includes instructions and a form for appealing the denial. Common reasons for denial include lack of medical necessity, out-of-network providers, missing documentation, incorrect billing codes, or multiple claims submitted.

Request Your Claim File (Before Filing an Appeal): A claim file contains information the insurer reviewed when deciding your claim, which can help you identify the specific issues to address in your appeal. Patients, advocates, and lawyers recommend requesting your claim file before you appeal. You can use ProPublica's claim file helper: https://projects.propublica.org/claimfile/.

The Appeals Process: Step by Step

1. Request an Explanation

Get a detailed reason for the denial in writing from the insurer. After extreme weather events, insurers may use generic denial codes that don't reflect the true circumstances.

2. Gather Documentation

Policy and Coverage Details: Get a copy of your insurance policy and confirm coverage terms, prior authorization rules, and medical necessity guidelines. You may need to get this from your employer or health insurer.

Medical Documentation: Collect doctor's notes, test results, lab work, prescription history, treatment plans, and referrals to prove medical necessity.

Communication Records: Keep a log of calls with your insurer and copies of related emails, letters, or faxes.

Appeal Forms and Letters: Obtain your insurer's appeal request form, a supporting letter from your doctor, and a personal appeal letter outlining your case with relevant medical evidence.

Regulatory Information: Research state insurance laws, ERISA rules (for employer-sponsored plans), and your rights for an external review.

Billing and Coding Details: To ensure you're being charged correctly, make sure the medical codes on your bill (CPT, ICD-10, and HCPCS) are accurate. You can request an itemized bill to carefully check for any mistakes or charges that don't match what you actually received.

- **Current Procedural Terminology (CPT):** These are codes used to describe medical procedures and services, like a doctor's visit, surgery, or lab test.

- **International Classification of Diseases (ICD-10):** These codes are used to describe diagnoses or health conditions, like flu or diabetes.

- **Healthcare Common Procedure Coding System (HCPCS):** These codes are used for services, equipment, and supplies not covered by CPT, like medical devices or ambulance services.

You can collect evidence connecting your health issue to a specific extreme weather event, including:

- ⚡ Weather reports for your location
- ⚡ Evacuation orders
- ⚡ Documentation of property damage
- ⚡ Medical records explicitly noting the connection to the weather event
- ⚡ Original claims
- ⚡ Correspondence related to the denied claim

3. Obtain a Support Letter from Your Physician

Request a detailed letter explaining the medical necessity of the treatment or service. This letter should reference specific medical records and, if applicable, include studies supporting the recommended treatment. For serious medical issues, your doctor can help gather clinical guidelines, peer-reviewed studies, or examples of similar approved cases to strengthen your appeal.

4. File an Internal Appeal

Submit your appeal within the timeframe specified by your insurer (typically 30 to 180 days from the denial notice). Include your gathered documentation. Provide a clear explanation of why the denial should be overturned. Include any additional information requested by the insurer. If delaying treatment could jeopardize your health, ask for an expedited appeal process.

5. Keep Records

Maintain copies of all communications. Document dates, times, and names of representatives. Record phone calls when possible. Retain copies of every document you send.

6. Pursue an External Review

If your internal appeal is denied, you have the right to an independent external review in most states. This process removes the final decision from the insurer's control. Visit: https://www.healthcare.gov/appeal-insurance-company-decision/external-review/. Be aware that regulations for external reviews vary by state.

Getting Help with Your Appeal

Patient Advocates

Patient advocates can improve your chances of a successful appeal if the process becomes overwhelming. These specialists have in-depth knowledge of complex medical billing and insurance policies. Organizations that can help you find a qualified advocate:

- **Patient Advocate Foundation (PAF)** • https://www.patientadvocate.org/

  A national nonprofit offering case management services for patients with chronic, life-threatening, and debilitating illnesses.

- **Greater National Advocates (GNA)** • https://gnanow.org/

  Provides an online directory of qualified healthcare advocates across the United States.

- **National Association of Healthcare Advocacy (NAHAC)** • https://nahac.com/directory-of-advocates#!directory

  Offers resources and support for finding professionals who can assist with insurance disputes.

Other Resources

- **State Health Insurance Assistance Programs (SHIPs)s** • https://www.shiphelp.org/

  Free, state-specific counseling for Medicare beneficiaries.

- **Hospital Patient Advocates**

  Many hospitals employ advocates or ombudsmen who can assist with billing issues and insurance appeals.

- **Centers for Medicare & Medicaid Services (CMS): No Surprises: Understand Your Rights Against Surprise Medical Bills • https://www.cms.gov/newsroom/fact-sheets/no-surprises-understand-your-rights-against-surprise-medical-bills**

  Provides an overview of patient rights and protections under the No Surprises Act, helping consumers avoid unexpected medical bills.

- **CMS: Key Protections Under the No Surprises Act • https://www.cms.gov/files/document/nsa-keyprotections.pdf**

  Outlines the main protections afforded by the No Surprises Act in a PDF format.

- **Hospital Pricing Files • https://www.hospitalpricingfiles.org/**

  Offers access to hospital prices, promoting transparency and enabling consumers to compare costs.

- **ProPublica: How Often Do Health Insurers Say No to Patients? No One Knows • https://www.propublica.org/article/how-often-do-health-insurers-deny-patients-claims**

  Investigates the frequency and reasons behind health insurance claim denials, highlighting the lack of transparency.

- **Vox: Never Pay a Medical Bill Without Asking These Questions First • https://www.vox.com/even-better/392930/medical-bill-debt-insurance-costs-pharmacy-payment-plan**

  What to ask while you're still at the hospital, doctor's office, or pharmacy.

- **Time: What to Do When Health Insurance Denies Care You Really Need • https://time.com/7014894/how-to-appeal-health-insurance-denial/**

  Provides guidance on steps to take when an insurance claim is denied, including how to appeal effectively.

- **ProPublica: Health Insurance Claim Denied? See What Insurers Said Behind the Scenes • https://www.propublica.org/article/find-out-why-health-insurance-claim-denied**

  Explores the internal processes of insurers regarding claim denials and how patients can access this information.

- **Patient Rights Advocate: How to Fight Medical Bill Overcharges • https://www.patientrightsadvocate.org/how-to-fight-medical-bill-overcharges**

  Provides strategies and steps to contest and reduce medical bill overcharges.

YOUR DOCTORS ARE
PARTNERS IN HELPING
YOU NAVIGATE AND
ADAPT TO THE HEALTH
RISKS POSED BY
CLIMATE CHANGE.
THAT PARTNERSHIP
BEGINS WITH A
SIMPLE BUT POWERFUL
QUESTION: *NOW THAT
I UNDERSTAND HOW
EXTREME WEATHER
AFFECTS MY HEALTH,
WHAT CAN I DO?*

TALKING TO YOUR DOCTORS ABOUT CLIMATE HEALTH

A Clinical Conversation on Climate and Health

Maybe you've been feeling unusually wiped out during heat waves. Or you're coughing more after a smoky week or sneezing non-stop when pollen counts spike. Maybe your child's asthma has been flaring up, or you've pulled more ticks off your dog (and your kids) than you ever have before.

Your doctors are partners in helping you navigate and adapt to the health risks posed by climate change. That partnership begins with a simple but powerful question: *Now that I understand how extreme weather affects my health, what can I do?*

To begin the conversation, ask targeted questions about your specific health conditions or medications. Below are examples organized by different climate-related events, designed to help you and your doctor safeguard your health in a changing world.

Heat

1. Are there any precautions I should take for my heart issues during heat waves?

2. Which of my medications could make me sun-sensitive? How do my combined medications interact with extreme heat? Does my psychiatric medication increase my sensitivity to heat or dehydration?

3. How can I keep my medications safe during extreme heat? What temperature range is safe for storing them?

4. How much water should I be drinking during heat waves?

5. How do I prepare for heat-related fatigue or dizziness that might mask hypoglycemia?

Air Quality

1. My allergies have been worse lately. Could this be related to rising pollen levels, and what can I do about it?

2. What should I do when my asthma flares up during wildfire smoke events or poor air quality days?

3. Should my inhaler use or dosage change during severe weather?

4. At what air quality index level should I start using my rescue medications?

5. How can I tell if my breathing problems are from weather changes versus an infection?

Heart Health

1. What signs should I watch for that my heart might be under strain?

2. Does air pollution or wildfire smoke affect my heart condition?

3. Do my blood pressure medications make me more sensitive to heat or dehydration?

4. Should I monitor my blood pressure more frequently during extreme weather?

5. What heart symptoms should I watch out for that require immediate emergency care during extreme weather?

Bugs

1. Lyme disease is everywhere. What symptoms should I watch for? What should I do if I think I have Lyme?

2. What are the treatments for illnesses spread by mosquitoes, like West Nile virus or Zika?

3. How do I know if diseases like dengue or chikungunya are a risk where I live?

4. Are vaccines or preventive medications available for any insect-borne illnesses I might encounter during travel?

5. What should I include in a travel health kit to manage insect bites and related illnesses?

Food and Water

1. What signs of food contamination or food-borne illness should I watch for?

2. How long is food safe in my refrigerator during a power outage?

3. What are the signs of waterborne illness I should watch for after flooding?

4. Should I get additional water testing done if there's been flooding in my area?

5. What precautions should I take when using tap water after a flood to ensure it's safe for drinking and cooking?

Medical Devices and Equipment

1. What should I do if my CPAP machine loses power?

2. How long will my medical device batteries last during an outage?

3. Are there non-electric alternatives for any of my medical devices?

4. Should I register my medical devices with emergency services?

5. What alternative treatments or therapies can I use if my medical device becomes inoperable during an outage?

Mental Health

1. How can I manage the stress and anxiety I feel during extreme weather?

2. Are there resources or counselors you recommend for managing climate-related anxiety?

3. How can I tell if my weather-related anxiety is becoming a more serious condition?

4. What coping techniques do you recommend during extended power outages or evacuations?

5. How can I address feelings of isolation during extreme weather if I can't leave home?

Special Needs

1. What additional precautions should older family members take?

2. How should I modify my pregnancy care plan during extreme weather?

3. What's the safest way to evacuate with limited mobility?

4. How do I ensure proper medication storage and dosing for my kids during extreme weather?

5. Are there specific shelters equipped to accommodate individuals with my disability during evacuations?

Long-Term Monitoring

1. What symptoms should I track to identify patterns of climate-related health issues?

2. How often should we review and update my health plan, given changing weather patterns?

3. What baseline measurements should I monitor and document?

4. Should I keep a weather diary alongside my symptom tracker?

5. Are there any new vaccinations recommended due to changing disease patterns?

Emergency Preparedness

1. Given my family's health conditions, what first aid should I include in my emergency plan?

2. How can I contact your office in the event of an emergency? When should I call emergency services versus waiting to reach your office?

3. How do I access treatment or medications if I lose power or need to evacuate?

4. How should I store medications, especially those requiring refrigeration, if there's a power outage? Can I obtain a backup supply of medicine in case the power goes out or if I need to evacuate?

5. How can I ensure my medical records are accessible during an evacuation? Which digital platforms or secure apps do you recommend for storing medical records that I can access offline?

Follow-Up Care

1. When should I schedule a follow-up appointment after being exposed to extreme weather?

2. What health screenings do you recommend after exposure to wildfire smoke, ash, or contaminated water?

3. How should I document any climate-related health impacts for my medical record?

4. How will we know if my treatment plan needs adjustment due to changing weather patterns?

Your Pediatrician

Climate poses unique risks to children, whose bodies and immune systems are still developing. Extreme weather, deteriorating air quality, and ecosystem shifts will impact your child's everyday health. Pediatricians can help interpret climate-related health risks based on your child's age, developmental stage, and preexisting conditions.

Heat and Dehydration

1. How can I keep my child safe from heat exhaustion or heat stroke during heat waves?

2. What are the signs of dehydration in children, and how can I address them quickly?

3. Should I limit or avoid activities for my children during extreme heat?

Air Quality

1. How can I manage asthma flare-ups during high pollen days or smoky air conditions?

2. Should my child's asthma medication plan be adjusted for severe weather events?

3. How can I tell the difference between seasonal allergies and something more serious?

Bugs

1. What symptoms should I watch for if a tick or mosquito bites my child?

2. What treatments are available for Lyme or West Nile virus?

3. What are the insect-borne diseases you are seeing in your patients?

Mental Health

1. How can I support my kids if they are experiencing anxiety or stress related to extreme weather or climate change?

2. What signs of emotional stress or trauma should I watch for in my child after a storm or flood?

3. Are there resources for counseling or therapy specifically for children coping with climate-related stress?

Emergency Preparedness

1. If we're evacuated, how can I ensure my child's asthma, allergies, or other health needs are managed?
2. What should I include in an emergency kit for my child's health needs?
3. Can you help me create a health-focused evacuation plan for my child? How can I ensure continuity of care for my child during evacuations or extended power outages?

Tips for a Productive Conversation

1. **Bring Records:** Prepare a detailed list of your child's medications, known allergies, and recent health concerns.
2. **Share your Observations:** Help your pediatrician understand your child's unique experiences with climate-related health effects. For example:

   a. *My child's eczema worsens on high-pollen days.*
   b. *My child feels faint after playing outside during heat waves.*

When Your Doctor Doesn't Have All the Answers

Climate is an evolving crisis, and its health impacts are advancing faster than medical systems can adapt. While your doctor may be a partner in your care, they might not always have immediate answers to your questions about climate-health risks. If you encounter a knowledge gap, don't lose hope. Here are some excellent options for clinicians to explore:

- Harvard T.H. Chan School of Public Health Toolkit for Clinicians • https://www.americares.org/what-we-do/community-health/climate-resilient-health-clinics/#toolkit

  This toolkit equips healthcare providers with the knowledge to understand and address climate-related health issues. It offers actionable strategies for protecting patients from heat stress, air quality issues, and infectious diseases linked to climate change. It includes case studies, patient education resources, and practical checklists that providers can use immediately in their practice.

- **American Medical Association (AMA)** • https://www.ama-assn.org/delivering-care/public-health/what-doctors-wish-patients-knew-about-climate-changes-health-impact

  The AMA has taken a strong stance on addressing the health impacts of climate change. This page focuses on what doctors wish patients knew about the link between climate and health. It's a good resource for encouraging your doctor to integrate climate-aware practices.

- **Enviromedics: The Impact of Climate Change on Human Health by Jay Lemery and Paul Auerbach**

  This book is written by two leading physicians in emergency and environmental medicine. It covers topics like the rise of heat-related illnesses, the spread of infectious diseases, and the mental health toll of climate stress. It's a great book to share with your doctor if they want to deepen their understanding of climate's medical implications.

- **American Academy of Pediatrics: "Climate Change and Children's Health: Building a Healthy Future for Every Child"** • https://publications.aap.org/pediatrics/article/153/3/e2023065505/196648/

  This article offers a comprehensive overview of how climate change affects children's health and what clinicians can do to mitigate these risks. It highlights topics such as:

  Children's unique vulnerability to extreme weather events, heat waves, and air pollution

  The importance of identifying climate-sensitive health conditions like asthma, allergies, and mental health in pediatric care

  Practical, evidence-based recommendations for integrating climate adaptation strategies into pediatric practice

- **AAP Resource Hub: "Disasters and Children"** • https://www.aap.org/en/patient-care/disasters-and-children

  The AAP has created a resource hub to address children's issues during climate-related disasters. This hub covers:

  Guidance for protecting children during hurricanes, wildfires, and floods.

  Pediatric-specific emergency preparedness tips, including access to life-sustaining medications and managing stress or trauma after disasters.

  Resources for pediatricians on how to educate families about risks and resilience strategies.

GLOSSARY

Air Quality Index (AQI): A daily rating system that indicates how safe the air is to breathe, based on measurements of primary pollutants including ground-level ozone, particulate matter, carbon monoxide, sulfur dioxide, and nitrogen dioxide. The scale ranges from 0 to 500, with higher values indicating greater pollution levels and associated health risks.

Alpha-Gal Syndrome (AGS): An allergic reaction to red meat triggered by tick bites, which can cause hives, digestive issues, and potentially anaphylaxis. In the U.S., it is linked to the lone star tick.

Anaplasmosis: A bacterial infection transmitted by ticks that causes fever, headache, muscle pain, and reduced white blood cell and platelet counts.

Bug-In: A survival strategy involving staying in place and sheltering within your home or another secure location during a disaster, focusing on fortifying your location with sufficient supplies.

Bug-Out: An evacuation strategy that involves leaving your home or current location in response to a disaster when staying in place becomes too dangerous, typically requiring a pre-packed emergency kit (a go bag).

County Animal Response Team (CART): Teams that operate at the county level are responsible for planning, coordinating, and executing animal-related disaster response efforts, including setting up temporary pet shelters and rescuing stranded animals.

Community Emergency Response Team (CERT): A program designed to train volunteers in basic disaster response skills such as fire safety, light search and rescue, team organization, and disaster medical operations.

Climate Health Literacy: Understanding how climate change affects health, identifying climate-related health risks, and adapting care strategies to protect against environmental threats.

Climate Literacy: The understanding of climate science, climate change impacts, and appropriate response strategies, enabling informed decision-making about climate adaptation and mitigation.

Combined Sewer Overflow (CSO): An event that happens when heavy rainfall exceeds the capacity of a combined sewer system. This system collects both stormwater and wastewater in the same pipes. When overwhelmed, the system can release untreated sewage into nearby waterways, polluting drinking water sources and increasing the risk of diseases.

Disaster Animal Response Team (DART): Broader teams that respond to emergencies involving animals at regional, state, or national levels, providing support in large-scale disasters with the rescue, sheltering, and care of impacted animals.

Disinfection Byproducts (DBPs): Toxic compounds formed when chemicals used to disinfect drinking water (like chlorine) react with natural substances in the water. Associated with an increased risk of bladder and colorectal cancers.

Disinformation: False information deliberately created and spread to mislead and confuse, often for political or economic gain; a form of propaganda.

Digital Health Literacy: The ability to use technology like electronic health records, health apps, digital platforms, and wearables to access and manage personal health information.

Emotional Freedom Techniques (EFT): A stress-reduction method that involves tapping on specific acupressure points while focusing on releasing stress-inducing issues, associated with lower cortisol levels.

Frostbite: A cold-related injury where body tissues freeze, most commonly affecting fingers, toes, nose, ears, cheeks, and chin. In severe cases, it can lead to permanent tissue damage and amputation.

Flash Flood: A rapid rise in water levels, typically occurring within 6 hours of the precipitation event that causes it, leaving little time for warning or evacuation.

Flood Warning: A warning that means that flooding is either imminent or actively occurring, and immediate action is necessary to ensure your safety.

Flood Watch: A warning that is issued when weather conditions suggest that flooding might occur.

Greenhouse Effect: The process by which gases in Earth's atmosphere, such as carbon dioxide and methane, trap heat from the sun, warming the planet. While natural, human activities have intensified this effect, leading to global warming.

Go Bag: A pre-packed emergency kit containing essential supplies (food, water, medications, documents, etc.) that can be quickly grabbed during evacuations.

Health Insurance Literacy: The ability to understand, navigate, and manage health coverage, including using online portals, understanding policy details, finding in-network providers, estimating costs, and navigating the claims process.

Health Literacy: The ability to read, understand, and act on health information, such as following medical advice, understanding prescriptions, and knowing when to seek care.

Heat Index: A measurement that combines air temperature and relative humidity to indicate how hot the weather feels to the human body, accounting for the reduced effectiveness of sweating in humid conditions.

Heat Wave: A period of abnormally hot weather compared to the expected conditions for an area and season, typically lasting three to seven days or more.

Hurricane Warning: Signals that hurricane conditions are expected within 36 hours.

Hurricane Watch: A warning that indicates hurricane conditions (sustained winds of 74 mph or higher) are possible within 48 hours.

Hypothermia: A dangerous condition that occurs when the body loses heat faster than it can generate it, causing the core temperature to drop to a dangerously low level, potentially leading to organ failure and death if untreated.

Independent Dispute Resolution (IDR): A process established by the No Surprises Act to mediate payment disputes between healthcare providers and insurers.

Media Literacy: The ability to analyze, evaluate, and understand media in all its forms, interpreting messages critically and revealing deeper processes and intentions.

Misinformation: False or inaccurate information, which may be shared without intention to deceive but can still cause harm.

Minimum Efficiency Reporting Value (MERV): The rating system used to measure the effectiveness of air filters in capturing particles of different sizes. Higher MERV ratings indicate better filtration.

No Surprises Act: A federal law enacted in 2022 designed to protect consumers from unexpected medical bills, particularly in emergency situations or when receiving care from out-of-network providers at in-network facilities.

NOAA Weather Radio: A nationwide network that broadcasts continuous weather information directly from National Weather Service offices, providing critical emergency alerts during disasters.

Particulate Matter (PM): Tiny solid and liquid particles in the air, classified by size (PM10, PM2.5, ultrafine particles). These particles can cause serious health problems when inhaled, particularly the smaller ones that can penetrate deep into the lungs and even enter the bloodstream.

Permethrin: A synthetic insecticide modeled after natural pyrethrins found in chrysanthemum flowers. It acts as both a repellent and insecticide, disrupting insects' nervous systems and causing paralysis and death upon contact.

PETS Act: The Pets Evacuation and Transportation Standards Act of 2006, which requires state and local emergency preparedness plans to account for the needs of individuals with household pets and service animals during major disasters or emergencies.

Polar Vortex: A vast, persistent whirl of low-pressure, frigid air that normally circles the Arctic (or Antarctic) but can wobble southward in winter, unleashing bursts of extreme cold far below the pole.

Polycrisis: A web of interconnected events, where multiple crises occur simultaneously and exacerbate one another, creating a chain reaction of problems that are more severe than they would be in isolation.

Post-Stabilization Care: Medical attention required after receiving emergency care, when a patient is stable but still needs additional treatment.

Solastalgia: The emotional distress caused by environmental changes that affect our sense of home or place. Homesickness experienced while still at home, as the environment around us deteriorates.

Specific Area Message Encoding (SAME): A technology used in NOAA Weather Radios that filters alerts by location, ensuring users receive only relevant emergency information for their area.

Urban Heat Island Effect: The phenomenon where urban areas experience significantly higher temperatures than surrounding rural areas due to heat-absorbing materials like concrete and asphalt, reduced vegetation, and heat generated by human activities.

Wet Bulb Globe Temperature (WBGT): A comprehensive measure of heat stress that incorporates air temperature, humidity, solar radiation, and wind speed to indicate how well the human body can cool itself through sweating.

Wireless Emergency Alerts (WEA): A public safety system that allows authorized government agencies to send geographically targeted emergency alerts to compatible mobile devices.

ADDITIONAL RESOURCES

Introduction to the Appendices:
Your Toolkit for Climate Health Preparedness

Throughout this book, we've explored the connections between climate change and human health. Our journey has taken us through sweltering heat waves, hazardous air quality, and devastating natural disasters. And the cascading effects these have on our physical, mental, and community well-being. We've examined how to communicate effectively in an era of misinformation, navigate healthcare systems during emergencies, and build resilient communities capable of facing climate threats together.

But understanding these issues is only the first step. The real-world scenarios we now face are messy, complicated, and always-changing. We need more than awareness. We need resources.

Information access has become increasingly vulnerable.

Websites disappear without warning, critical data becomes unavailable during emergencies, and misinformation spreads at alarming rates. The physical infrastructure we often take for granted (electricity, clean water, internet connectivity, and access to healthcare) can all fail during climate disasters.

These appendices help transform your climate-health literacy into action. They bridge the gap between knowing about climate threats, to effectively responding to them when they occur.

Each appendix connects directly to the core themes and chapters of this book:

Extreme Weather Events

- ⚡ Appendix A: Flood Resources
- ⚡ Appendix B: Power Outage Resources
- ⚡ Appendix C: Extreme Cold Resources

How to Be Prepared

- ⚡ Appendix D: Resources for Emergency Preparedness

Secondary Effects of Extreme Weather

Communication and Media

Navigating Healthcare

EXTREME WEATHER EVENTS

Appendix A: Flood Resources

Homeowners' insurance typically does not cover flood damage. The catastrophic flooding in North Carolina caused by Hurricane Helene in September 2024 highlighted this. Despite the state's susceptibility to hurricanes and flooding, only about 2.6 percent of North Carolina properties were covered by flood insurance at the time of the hurricane. The low coverage rate left many homeowners unprotected against the extensive flood damage caused by the storm.

The National Flood Insurance Program (NFIP), administered by FEMA, provides flood insurance to homeowners in participating communities. However, the program has faced criticism for outdated flood risk maps, which led to homeowners underestimating their flood risk. The program's focus has been on coastal and river flooding. Given the increasing frequency and intensity of flooding events, homeowners should consider purchasing flood insurance, even if they are not located in designated high-risk areas.

Assessing your property's flood risk, understanding the limitations of standard insurance policies, and exploring coverage options through the NFIP or private insurers is crucial to preparing for future floods. Climate change is making flooding more frequent, more intense, and harder to predict. But with the right tools, you can understand your risks and take informed action.

To prepare for flood events, you first need to know your exposure. Floods don't follow the rules. They can happen with or without rain, upstream or downstream, from hurricanes or broken infrastructure. Understanding where water flows, and where it backs up, is as important as tracking the weather.

Assessing Long-Term Risk

The following sites can help you assess the long-term risk to your home, neighborhood, or town.

- **First Street Foundation • https://firststreet.org/**

  A powerful tool offering a 30-year risk outlook for your property. It provides a 1 to 10 rating for flood, wildfire, extreme heat, and wind. Great for real estate decisions, home upgrades, and local planning.

- **ClimateCheck • https://climatecheck.com/**

  Get a free climate risk report by entering an address. You'll receive a personalized score for flood, heat, drought, and wildfire—plus insights into how these threats may worsen over time.

- **FEMA National Risk Index • https://hazards.fema.gov/nri/**

  This tool assesses national hazard risk using a combination of exposure, social vulnerability, and resilience metrics.

- **FEMA Flood Map Service Center • https://msc.fema.gov/portal/home**

  If you're buying a home or renewing insurance, this is your go-to source for official floodplain maps. Search by address to see if you're in a Special Flood Hazard Area.

- **NOAA Coastal Flood Exposure Mapper • https://coast.noaa.gov/floodexposure/**

  Visualizes coastal flood threats from sea-level rise, hurricanes, and storm surge. Extremely helpful for town hall planning meetings or local adaptation projects.

- **NASA Flooding Analysis Tool • https://sealevel.nasa.gov/flooding-analysis-tool/projected-flooding?**

  Forecasts high-tide flooding based on lunar and sea-level data. Especially valuable for coastal communities facing regular nuisance flooding.

- **Climate Central Surging Seas Risk Zone Map • https://sealevel.climatecentral.org/**

  See which parts of your city or region are most threatened by sea-level rise. This tool uses peer-reviewed climate data to project future flood scenarios.

- **NOAA National Water Prediction Service • https://www.weather.gov/owp**

  Stay updated with real-time streamflow and water predictions. A key resource during rainfall events, snowmelt, or potential dam breaches.

- NOAA River Forecast Centers • https://water.noaa.gov/about/rfc

  Track current river conditions, forecasted crest levels, and flood stage alerts. This helps residents and responders make critical decisions early.

Surfacing Invisible Flood Risks

Floodwaters are rarely clean. They can stir up hidden toxic sites and contaminate water supplies, particularly in industrial or low-lying areas. Floods don't hit all communities equally. Many low-income areas are located near industrial sites or downstream infrastructure, making recovery harder and more dangerous. The following tools help surface invisible risks.

- EPA Superfund Site Info • https://www.epa.gov/superfund

  Discover if you live near a Superfund site and how floodwaters might spread toxic waste. Each site page details cleanup status, health concerns, and risk levels.

- EPA Envirofacts Database • https://enviro.epa.gov/

  Search your zip code or county to learn about air, water, and soil pollutants in your area. Useful for people with health concerns or kids at home.

- EPA Toxics Release Inventory • https://www.epa.gov/toxics-release-inventory-tri-program

  Monitors chemical releases from factories and industrial sites. It's one of the best sources to see who's polluting—and how much.

- EPA Community Impact Map • https://www.epa.gov/community-port-collaboration/community-port-collaboration-toolkit

  Maps the overlap between toxic sites and climate risks, such as floods and storm surges. Helps visualize the compound threats some neighborhoods face.

Further Resources

Whether you need flood insurance, an emergency kit, or post-flood cleanup guidance, the following tools will help.

- National Flood Insurance Program (NFIP) • https://www.fema.gov/flood-insurance

  Flood damage is not covered by most homeowners insurance.

- **Ready.gov Flood Resource Center** • **https://www.ready.gov/floods**

  Information on emergency kits, evacuation planning, and education on flash flood safety.

- **CDC Guidelines for Cleaning Safely After a Disaster** • **https://www.cdc.gov/natural-disasters/ safety/?**

  Floods leave behind more than mud. Learn how to safely reenter your home, check for mold, and handle contaminated surfaces.

- **TruePrepper Flood Survival Kit Guide** • **https://trueprepper.com/flood-survival-kit/**

  A clear, practical guide to building your flood survival kit. Covers gear, supplies, and items people often forget, like backup documents and protective gloves.

Appendix B: Power Outage Resources

- **Heating and Cooling Center Finder** • https://www.211.org/

  Tool to locate local heating and cooling centers during extreme weather events, helping individuals find relief during emergencies and ensuring safety during heat waves or severe cold spells.

- **FEMA Mobile App** • https://www.fema.gov/about/news-multimedia/mobile-products

  Official FEMA mobile app providing real-time emergency alerts, preparedness checklists, disaster recovery tips, customizable alerts, safety guidance, and detailed resources to assist with disaster readiness and response.

- **Red Cross Shelter Finder** • https://www.redcross.org/get-help/disaster-relief-and-recovery-services/find-an-open-shelter.html

  Resource for locating open emergency shelters during disasters or emergencies, providing real-time shelter availability and essential information to individuals displaced by natural disasters or other emergencies.

- **Cooling Centers by State** • https://nchh.org/information-and-evidence/learn-about-healthy-housing/emergencies/extreme-heat/cooling-centers-by-state/

  National Center for Healthy Housing directory listing cooling centers available across the United States.

- **Best Home Generators** • https://www.thespruce.com/best-home-generators-4173501

  Guide to home generators, offering in-depth recommendations tailored to various household needs, power capacities, fuel types, portability options, and budget constraints.

- **Power Outages** • https://www.ready.gov/power-outages

  Official government guidelines offering steps for preparing for and safely navigating power outages, including emergency kit recommendations, communication planning, safety protocols, and information on how to effectively manage prolonged power disruptions.

- **Tools for a Power Outage** • https://www.nytimes.com/wirecutter/reviews/tools-for-a-power-outage/

  Curated list and recommendations of essential tools and supplies necessary for effectively managing and recovering from power outages, covering everything from lighting solutions, backup power options, portable battery packs, and emergency radios.

- **PowerOutage.us • https://poweroutage.us/**

  Interactive map of power outages across the United States, offering real-time updates, statistics, and detailed outage information by state and utility company.

- **American Red Cross Power Outage Safety • https://www.redcross.org/get-help/how-to-prepare-for-emergencies/types-of-emergencies/power-outage.html**

  Guidelines on safety practices, emergency preparedness, and recovery strategies to effectively manage and safely navigate power outages.

Appendix C: Extreme Cold Resources

- **National Weather Service Social Media: Winter & Cold** • https://www.weather.gov/wrn/winter-storms-sm

  Tips and information to stay safe before, during, and after a winter storm.

- **National Weather Service Cold Weather Safety** • https://www.weather.gov/safety/cold-links

  Resources on extreme cold, hypothermia, frostbite, and cold weather preparedness.

- **CDC Winter Storm Safety** • https://www.cdc.gov/winter-weather/safety/stay-safe-during-after-a-winter-storm-safety.html

  Guidance on staying safe during and after a winter storm, including heating, driving, and power outage safety.

- **CDC Winter Weather Prevention** • https://www.cdc.gov/winter-weather/prevention/index.html

  Tips to prevent cold-related health risks, including dressing for the cold, preparing your home, and vehicle safety.

- **CDC Frostbite Guide** • https://www.cdc.gov/winter-weather/media/pdf/avoid-spot-treat-frostbite.pdf

  How to avoid, spot, and treat frostbite during extreme cold conditions.

HOW TO BE PREPARED

Appendix D: Resources for Emergency Preparedness

- **Best Gear for Your Bug-Out Bag** •
  https://www.nytimes.com/wirecutter/reviews/best-gear-for-your-bug-out-bag/

  A curated guide to the best gear for assembling a bug-out bag, ensuring you're ready to evacuate quickly with essential tools and supplies. Includes expert-tested recommendations for portable water filters, multi-tools, emergency radios, and more.

- **Emergency Preparedness Gear Reviews** •
  https://www.nytimes.com/wirecutter/emergency-preparedness/

  A comprehensive overview of all emergency preparedness resources, featuring expert analysis and testing of equipment for various disaster scenarios. Covers everything from emergency food supplies to backup generators.

- **Emergency Preparedness Essentials** •
  https://www.nytimes.com/wirecutter/reviews/emergency-preparedness/

  Focused on preparing for a range of emergencies, this guide helps you build a customized emergency kit based on your location, living situation, and specific risks. Includes recommendations for food storage, water purification, and communication tools.

- **Reddit Preppers Wiki** • https://www.reddit.com/r/preppers/wiki/index

  A community-driven knowledge base offering practical tips, gear suggestions, and real-life advice on prepping for disasters. Includes sections on financial preparedness, food storage techniques, and first-hand accounts from disaster survivors.

- **Reddit Preppers Subreddit** • https://www.reddit.com/r/preppers/

  Active community forum where preppers of all experience levels share insights, ask questions, and discuss emerging threats and preparation strategies. Features regular threads on budget prepping and region-specific concerns.

- **Coursera Disaster Preparedness** • https://www.coursera.org/learn/disaster-preparedness

  An online course offering foundational knowledge about disaster planning, risk reduction, and emergency response. Taught by experts in emergency management and public health, with practical assignments to build your disaster readiness skills.

- **The Urban Prepper YouTube Channel** • https://www.youtube.com/c/TheUrbanPrepper

  A video resource for urban emergency preparedness, with tips and tutorials on surviving in city environments during disasters. Features demonstrations of gear testing, apartment-friendly prep solutions, and urban evacuation strategies.

- **The Prepared** • https://theprepared.com/

  An in-depth guide for all levels of emergency preparation, covering everything from first aid to gear reviews. Features evidence-based recommendations and a no-fear-mongering approach to rational preparedness.

- **Primal Survivor** • https://www.primalsurvivor.net/

  A resource focusing on survival skills, gear, and practical advice for emergencies and outdoor survival. Includes detailed guides on wilderness survival techniques, food preservation, and off-grid living strategies.

- **Apartment Prepper** • https://apartmentprepper.com/

  Specialized advice for urban dwellers with limited space, focusing on preparation methods for those living in apartments or small homes. Includes creative storage solutions and building community resilience in urban settings.

- **Creek Stewart** • https://www.creekstewart.com home

  A comprehensive resource from survival expert and author of The Disaster Ready Home, Creek Stewart. Featuring wilderness survival skills, disaster preparedness guides, and self-reliance techniques, the site offers instructional videos, gear recommendations, and information about Creek's courses. His approach emphasizes practical, field-tested methods for emergency situations and everyday readiness.

Appendix E: Your Go Bag

EMERGENCY COMMUNICATION	Radio	Battery-powered or hand-crank radio with NOAA Weather Radio capability for emergency alerts.
	Cell phones	Cell phones with multiple charging options: wall charger, car charger, portable battery pack, and solar/hand-crank backup charger.
LIGHTING	Flashlights, headlamps, lanterns	LED flashlights with extra batteries, or hand-crank/solar-powered flashlights that don't require batteries.
		LED headlamps for hands-free lighting during tasks.
		Battery-powered LED lanterns for room lighting during extended outages.
FOOD AND WATER	Food	Non-perishable food for several days. Canned goods, grains, snacks, dried meat and fruits, and other shelf-stable items you enjoy.
	Water	One gallon per person per day for drinking and sanitation. FEMA recommends a minimum three-day water supply, but a 14-day supply is ideal in high-risk areas.
	Water purification options	Purification tablets or systems for clean water.
	Infant supplies or baby food	Formula, bottles, diapers, wipes, and diaper rash cream.
	Manual can opener and utensils	For opening canned food.
HEALTH AND HYGIENE	First Aid Kit and medical supplies	Comprehensive kit with basic medical supplies. Pain relievers, anti-diarrhea, antacids, and laxatives. Soap, hand sanitizer, disinfecting wipes. Feminine supplies and other personal hygiene items. Moist towelettes, hand sanitizer, garbage bags, and plastic ties.
	Diapers	Diapers for infants or toddlers.
	Medications	30-day supply of prescription medications and other essentials.
	N95 masks	Masks for respiratory protection.
	Gloves	Work gloves for handling rough items.
SAFETY AND PROTECTION	Masks	For contaminated air, important for wildfires or hazardous conditions.
	Fire	Fire extinguisher, waterproof matches. Fire-starting tools.
	Whistle	A loud whistle to attract attention if you need help.
	Reflective triangles, flairs	Warning triangles for roadside emergencies. Emergency signaling flares.

POWER	Portable power	Portable electric or solar power as an alternative to gas generators. Can power small appliances without emitting exhaust.
SAFETY AND PROTECTION	Plastic sheeting and duct tape	To shelter in place if necessary. Versatile tool for securing and repairing items.
	Poncho, survival blankets	Waterproof poncho for rain protection. Compact blankets for warmth.
DOCUMENTS AND FINANCES	Essential Documents	Copies of important documents in a waterproof container. Insurance policies, identification, passports, and financial records. Physical copies are important if you are unable to connect to the internet. Cash or traveler's checks in case of card or system failure.
	Password tool	To safely encrypt passwords.
TOOLS AND EQUIPMENT	Tools	Wrench, pliers, multi-tool, knife, and shovel. A gas shut-off tool is recommended for safely turning off utilities in an emergency.
	Jack and lug wrench	For tire changing.
	Rope	For tying or securing items.
	Rock salt	Salt for winter road conditions.
WARMTH AND COMFORT	Comfort items	Sleeping bags, warm blankets, a change of clothes, and sturdy shoes.
NAVIGATION	Local maps	Used for navigation if electronic systems fail.
PETS	Pet supplies	Food and extra water for pets. See Chapter on Emergency Preparedness for Pets for a detailed list.

Appendix F: Emergency Resources for Pet Owners

- **CDC Pet Safety in Emergencies •**
  https://www.cdc.gov/healthy-pets/emergency-preparedness/index.html

  Information from the Centers for Disease Control and Prevention about keeping pets safe during various types of emergencies.

- **Ready.gov Prepare Your Pets for Disasters • https://www.ready.gov/pets**

  Official government guidance on pet preparedness, including evacuation planning, emergency kits, and special considerations for different types of disasters.

- **American Red Cross Pet Disaster Preparedness • https://www.redcross.org/get-help/**
  how-to-prepare-for-emergencies/pet-disaster-preparedness.html

  Resources and guidance from the American Red Cross on including pets in your family emergency plan.

- **ASPCA Disaster Preparedness for Pets • https://www.aspca.org/pet-care/general-pet-care/**
  disaster-preparedness

  The American Society for the Prevention of Cruelty to Animals offers advice on preparing for and managing pets during disasters.

- **National Alliance of State Animal and Agricultural Emergency Programs (NASAAEP) •**
  https://www.thenasaaep.com/best-practices

  Current best practices in animal emergency management from a consortium of state-level animal and agricultural emergency programs.

- **AVMA First Aid Tips for Pet Owners • https://www.avma.org/resources-tools/pet-owners/**
  emergencycare/first-aid-tips-pet-owners

  Comprehensive advice from the American Veterinary Medical Association on providing first aid for pets during emergencies.

Appendix G: Phone Emergency Features

Apple (iOS) Emergency Features

iPhones come equipped with several emergency features. The Emergency SOS function can be activated by simultaneously pressing and holding the side button and one of the volume buttons until the Emergency SOS slider appears on screen.

On newer iPhone models (iPhone 14 and later, with iOS 16.1+), there's also Emergency SOS via satellite that works in remote locations without cellular or Wi-Fi coverage by following on-screen instructions to connect to a satellite.

The Medical ID feature, configurable through the Health app, allows first responders to access medical information even when your phone is locked. To enable this, open the Health app, tap your profile picture, select Medical ID, press Edit, and ensure the "Show When Locked" option is enabled. When you use Emergency SOS, your iPhone can automatically share your Medical ID information with emergency services in supported regions.

Apple's Find My iPhone service, accessible through the Find My app, serves a dual purpose: it helps locate your device if it goes missing and enables location sharing with trusted contacts.

To share your location with friends or family members, open the Find My app, tap the "Me" tab, and toggle on the "Share My Location" option.

iPhones are configured to receive *Wireless Emergency Alerts (WEA)* for severe weather events and other public safety warnings. You can manage these notifications by navigating to Settings, selecting Notifications, and scrolling down to the "Government Alerts" section.

Android Emergency Features

Android devices offer similar emergency capabilities, though this varies by manufacturer. On most Android phones running Android 12 or later, the Emergency SOS function can be accessed through the Settings app under "Safety and Emergency." The feature is typically activated by quickly pressing the power button five times in succession, though on Samsung devices it may be three or four presses. This action can automatically trigger an alarm sound, call emergency services, and send SOS messages with your current location to designated emergency contacts.

Android's Emergency Information feature allows you to store critical medical details and emergency contact information accessible from the lock screen. First responders can access this information without unlocking the device. To set this up, go to Settings, select Safety and Emergency, and then Medical Information.

For location sharing, you can use Google Maps or the Emergency SOS feature itself. When Emergency SOS is triggered, it can be configured to automatically share your location with emergency contacts.

For manual location sharing through Google Maps, open the app, tap your profile picture, select Location sharing, and choose Share Location. Your emergency contacts will receive updates about your current location for a specified period.

Android devices are equipped to receive Wireless Emergency Alerts for critical situations. To ensure these alerts are enabled, navigate to Settings, select Notifications, and then Wireless Emergency Alerts. Additionally, Emergency Location Service (ELS) automatically provides precise GPS coordinates to emergency services when you call or text an emergency number, potentially reducing response times.

A useful feature for emergency preparedness is Google Maps' offline functionality. You can download maps for offline use, which proves invaluable during power outages or in areas with poor cellular coverage. To download maps, open Google Maps, tap your profile picture, select Offline maps, choose Select your own map, and download the desired area.

SECONDARY EFFECTS OF EXTREME WEATHER

Appendix H: Resources for Parents

- NASA Climate Kids • climatekids.nasa.gov

  Child-friendly platform with games, activities, and educational resources from NASA.

- Harvard C-CHANGE: Kids and Climate • https://hsph.harvard.edu/research/climate-health-c-change/climate-kids-and-health/

  Resources focused on children's health and climate education.

- TED Ideas: Tips on Talking to Kids About Climate Change • https://ideas.ted.com/tips-on-how-to-talk-to-kids-about-climate-change/

  Expert tips for sensitive climate discussions with kids.

- CLEAN Network • https://cleanet.org/index.html

  Free teaching resources for K–12 through college. A community of over 800 members committed to improving climate and energy literacy for all.

- NASA's Global Precipitation Measurement (GPM) Education • https://gpm.nasa.gov/education/

  Free materials and activities about weather, climate, and precipitation.

- Building Resilience in Children Through Coping Skills • https://www.redcross.org/get-help/how-to-prepare-for-emergencies/teaching-kids-about-emergency-preparedness/building-resilience-in-children-through-coping-skills.html

  Red Cross guidelines for building emotional resilience in children.

- SafeWise: Preparing Kids for Emergencies • https://www.safewise.com/how-to-prepare-kids-for-emergencies/

  Tips to teach children safety and proper reactions during emergencies.

- Talking to Children About Disasters • https://www.healthychildren.org/English/healthy-living/emotional-wellness/Pages/Talking-to-Children-about-Disasters.aspx

  Guide on emotional and practical disaster discussions.

- Managing Climate Anxiety as a Parent • https://www.popsugar.com/parenting/managing-climate-anxiety-parent-49355515

  Tips for managing parental climate anxiety while supporting children.

COMMUNICATION AND MEDIA

Appendix I: Information and Disinformation Resources

Climate Change and Environmental Information

- **Climate Central** • https://www.climatecentral.org/resources

  Authoritative research center providing interactive tools, maps, and analysis of climate data and trends.

- **Climate Desk** • https://www.climatedesk.org/

  Collaborative journalism partnership exploring climate change's societal, political, and economic implications.

- **EcoWatch** • https://www.ecowatch.com/

  Environmental news platform covering sustainability innovations, conservation efforts, and ecological advocacy.

Fact-Checking and Misinformation Tools

- **BBC Reality Check** • https://www.bbc.com/news/reality_check

  BBC's dedicated verification service examining factual claims in public discourse and media.

- **FactCheck.org** • https://www.factcheck.org

  University of Pennsylvania project monitoring accuracy in political statements and campaign rhetoric.

- **Snopes** • https://www.snopes.com

  Pioneering fact-checking website investigating urban legends, viral content, and widespread misconceptions.

- **Skeptical Science** • https://www.skepticalscience.com

  Scientific resource examining climate claims against peer-reviewed research evidence.

- **DeSmog** • https://www.desmog.com

  Investigative journalism outlet exposing corporate-funded climate denial campaigns and disinformation networks.

Health Information and Medical Resources

- **CDC Primary** • https://www.cdc.gov/index.html

  Official U.S. public health agency providing authoritative disease prevention guidelines and epidemiological data.

- **Backup of CDC** • https://restoredcdc.org/www.cdc.gov/

  RestoredCDC.org is an independent project that archived the CDC website on January 6, 2025.

- **MedlinePlus** • https://medlineplus.gov/

  Consumer-friendly medical encyclopedia offering reliable, jargon-free health information from the U.S. National Library of Medicine.

- **National Institutes of Health (NIH)** • https://www.nih.gov/health-information

  Premier U.S. biomedical research institution publishing evidence-based health resources and clinical trial opportunities.

- **WebMD** • https://www.webmd.com/

  Comprehensive health information portal featuring symptom checkers, treatment guides, and medical news.

- **PubMed** • https://pubmed.ncbi.nlm.nih.gov/

  Extensive searchable database of biomedical literature, clinical studies, and scientific research abstracts.

- **Academic Research Google Scholar** • https://scholar.google.com/

  Specialized search engine indexing scholarly literature across disciplines and publication formats.

Political Influence and Financial Transparency

- **OpenSecrets.org** • https://www.opensecrets.org/

  Nonpartisan research organization tracking money flows in U.S. politics and their policy impacts.

- **ProPublica Nonprofit Explorer** • https://projects.propublica.org/nonprofits/

  Investigative tool revealing financial activities of charitable organizations through IRS records.

- LobbyView • https://www.lobbyview.org/

  Database monitoring corporate influence through congressional lobbying activities and expenditures.

- Facebook Ad Library • https://www.facebook.com/ads/library/

  Transparency tool revealing sponsors, spending, and targeting behind political and issue advertisements.

Media Bias and News

- AllSides • https://www.allsides.com

  News aggregator presenting multiple political perspectives on each story with bias ratings.

- Media Bias / Fact Check • https://mediabiasfactcheck.com/

  Independent media evaluation service rating news sources on factual reporting and political orientation.

- Feedly • https://feedly.com/news-reader

  AI-powered news curation platform organizing content from diverse sources into personalized feeds.

- Flipboard • https://flipboard.com/

  Magazine-style content discovery app aggregating news and articles based on user interests.

- Future of Media Project • https://projects.iq.harvard.edu/futureofmedia/index-us-mainstream-media-ownership

  Harvard research initiative mapping ownership concentration in American media landscapes.

- The Conversation • https://theconversation.com/us/environment

  Platform publishing analysis and commentary from academic experts on environmental issues and climate science.

Tools and Training for Identifying Misinformation

- Bad News • https://www.getbadnews.com

  Interactive game simulating how misinformation spreads online, teaching players to recognize manipulation tactics.

- **Cranky Uncle** • https://crankyuncle.com

  Educational app using gamification to help users identify and counter common logical fallacies in misleading arguments.

- **Propaganda Critic** • https://propagandacritic.com

  Comprehensive resource analyzing historical and contemporary propaganda techniques with real-world examples.

- **News Literacy Project** • https://newslit.org/

  Educational initiative offering classroom resources and public programs to help citizens discern fact from fiction.

- **Deepware Scanner** • https://www.deepware.ai

  AI-powered tool identifying manipulated videos and deepfakes through advanced detection algorithms.

NAVIGATING HEALTHCARE

Appendix J: Climate Health Emergency Preparedness Assessment

This checklist is designed to help you prepare for extreme weather, especially if you or someone in your family is managing a health condition.

General Emergency Preparedness

1. Do you have a written emergency plan for power outages, extreme weather, or evacuation? Have you shared this plan with your family or caregivers?

2. Are you familiar with the emergency shelters in your local area? Are they accessible? Have you verified their status? Were they open during past emergencies?

3. Do you have a reliable backup power source for medical devices (generators or batteries)? Have you recently tested these power sources to ensure they work properly?

4. Do you know about the community resources or programs in your area that provide assistance during emergencies? Do you know how to contact them in a crisis?

5. Have you prepared a go bag with essential medical and personal supplies? Does it include items for your entire family, such as prescriptions or special equipment?

6. Have you coordinated with your child's school regarding emergency planning?

7. Are there backup transportation or pickup arrangements in place for your children in the event of an emergency?

Medication and Equipment

1. Do you have a three-month supply of essential medications, including backup prescriptions? Have you considered how to manage this if supply chains are disrupted?

2. Are your medications stored in a way that protects them from extreme temperatures or power outages? Do you have alternative storage solutions like portable coolers for temperature-sensitive medications? Have you tested these to make sure they work?

3. Do you know which pharmacies or supply centers remain open during disasters? Do you have their contact information?

4. Do you have backup medical equipment and spare parts?

5. Do you know how to maintain your medical equipment, and do you perform regular checks?

Transportation and Access

1. Do you have a plan for transportation to medical facilities during emergencies? Does this plan account for road closures or public transportation disruptions?

2. Are you familiar with multiple treatment centers or pharmacies in case your usual location is inaccessible? Have you mapped out alternative routes?

3. Have you coordinated with caregivers or community services to assist with evacuation or transportation needs? Do they have a copy of your emergency plan?

4. Have you ensured your car is always fueled and equipped with basic emergency supplies?

5. Do you have an emergency plan in place for your pets, including food, medications, and shelter?

6. Are your pets microchipped, and do they have updated identification?

Air Quality and Flooding

1. Have you identified safe spaces with good air quality during wildfire seasons or pollution events? Do you know their locations and how to access them?

2. Do you have a home air filtration system or a portable air purifier? Are they properly maintained and functional?

3. Do you have masks or other protective gear for days with poor air quality? Do you have enough supplies for all family members?

4. Do you check your apps for local air quality?

5. Have you assessed your home's vulnerability to flooding, hurricanes, or other disasters?

6. Do you have flood-proof containers or storage for your supplies?

Food and Water

1. Do you have a supply of clean drinking water and food? Have you calculated how much water and food you need for an extended emergency?

2. Have you identified allergen-free food options or special diet needs in your emergency plan? Are these items easily transportable?

3. Do you have access to water purification in case of water supply disruptions? Have you tested these methods to ensure they work?

4. Are you familiar with your local resources for obtaining clean water during an emergency?

Communication and Documentation

1. Do you have updated copies of medical records, including recent lab results? Are these documents stored securely but accessible in an emergency?

2. Have you clearly labeled emergency contacts and medical conditions on a medical alert system or bracelet? Are these labels up to date?

3. Is your emergency plan shared with family members, neighbors, or caregivers? Do they understand their roles in executing the strategy?

4. Do you have access to emergency communication tools, such as a charged cell phone or NOAA radio? Are these tools included in your go bag?

5. Have you practiced using communication tools with your family members?

6. If English is not your first language, are emergency instructions and medical alerts available in your preferred language?

7. Have you included cultural or religious considerations in your emergency plan?

Stress and Mental Health

1. Have you identified ways to manage stress or anxiety during emergencies? Do you have access to relaxation techniques, apps, or resources?

2. Are you familiar with local mental health support services or hotlines? Do you have their contact information?

3. Do you have a plan for maintaining mental health treatment during disruptions? Does this plan include telehealth?

4. Have you discussed your mental health needs with trusted friends or family members who can provide support?

5. Have you participated in community workshops or training sessions on stress management during emergencies?

Special Considerations

1. Do you have specific needs related to mobility, hearing, or vision that require additional planning? Have you accounted for these in your evacuation plan?

2. Do you have specialized equipment or supplies for newborns, children, or seniors in your household? Are these items included in your emergency kit?

3. Have you practiced evacuation drills or other emergency scenarios with your family? Do these drills cover all likely scenarios?

4. Have you ensured that all assistive devices are functional and have spare parts ready?

5. Have you included sensory or behavioral aids for individuals with cognitive needs in your emergency plan?

6. Are caregivers trained to handle emergencies for individuals with developmental disabilities?

Additional Preparedness Steps

1. Have you reviewed your plan with your healthcare provider to ensure it meets all your family's medical needs? Have you scheduled regular check-ins to update the plan?

2. Are you signed up for local emergency alerts or notification systems? Have you tested these systems to ensure you receive alerts?

3. Do you have insurance coverage for medical needs during disasters? Have you reviewed your policy to understand what's covered?

4. Are you aware of financial assistance programs available for healthcare disaster recovery? Have you gathered the necessary documents to apply quickly if needed?

5. Have you participated in community workshops or training sessions on emergency preparedness? Are you familiar with local initiatives?

6. Do you have plans for managing prolonged power or water outages?

7. Have you considered solar power or other renewable energy options as backups?

8. Do you have a financial contingency plan for unexpected costs during emergencies?

9. Have you set aside funds for emergency supplies or evacuation needs?

Appendix K: 2X Medicare Disclaimer

Healthcare billing can be incredibly confusing and stressful, especially when facing unexpected or high medical charges. Marshall Allen's "2X Medicare Rate Disclaimer" is a strategy to help protect you from unfairly high medical bills. Marshall Allen's advice is: *Everything is negotiable in healthcare billing.* Your disclaimer starts that negotiation, and having it gives you leverage.

What Is the 2X Medicare Rate Disclaimer?

This disclaimer is a simple sentence that you can add to medical consent or financial responsibility forms: "I consent to appropriate treatment and, including applicable insurance payments, agree to be responsible for reasonable charges up to twice the Medicare rate." You can even add the additional qualifier at the end, "following receipt of an itemized bill for appropriate treatment coded at the correct Level." By adding this, you are agreeing to pay a fair price for care (up to twice what Medicare pays), thereby preventing providers from billing excessive charges.

When Should You Use the Disclaimer?

Use this anytime you sign medical paperwork, especially when:

- ⚡ You're in an emergency room or urgent care, which may have out-of-network providers.

- ⚡ You visit a specialist who is out-of-network or performs tests, since your insurance may not fully cover the costs.

- ⚡ You receive elective procedures at outpatient surgery centers or imaging facilities.

How Do You Use the Disclaimer?

1. Carefully check the form. Look for language stating that you must pay all charges not covered by your insurance.

2. Cross out unfair language. Draw a line through phrases like "I agree to pay all charges not covered by insurance."

3. Clearly write the disclaimer sentence next to your crossed-out section. Initial this change.

4. Take a photo or request a copy. Always keep a copy of the modified form for your records, which helps protect you later.

If the forms are digital, request a paper version or submit a written note separately stating the disclaimer. Then, email it to the billing department as soon as possible.

Avoid Overpaying: A Simple Guide Using Medicare Rates

Step 1: Research Before Treatment
Visit https://www.medicare.gov/procedure-price-lookup/ before your appointment. Enter your zip code and search for the specific procedures or services you'll receive. This gives you the baseline Medicare rate for your area.

Step 2: Calculate Your Maximum Liability
Take the Medicare rate and multiply by two. This is the absolute maximum you should pay under your consent agreement. For example, if Medicare pays $200 for a procedure, your maximum charge should be $400.

Step 3: Document Everything
Print or screenshot the Medicare rates for your records. Keep this information with your consent form and any bills you receive.

Challenge Excessive Charges

Contact the billing department if charges exceed twice the Medicare rate. Explain your signed agreement clearly and politely. You might say: "I agreed to pay up to twice the Medicare rate, as documented on my signed intake form. Let's adjust the bill accordingly."

If You Face Pushback

If providers or billing staff resist your disclaimer, calmly explain that you're willing to pay fair rates (twice the Medicare rate is widely considered reasonable). Changing contract terms before signing is your right. Once providers deliver care after you've modified terms, they've essentially accepted your terms.

If you're facing resistance during a billing or insurance conversation, politely ask to speak with a supervisor. You can request to speak with a patient advocate or a billing manager at the facility. Remember, under federal law (EMTALA, the Emergency Medical Treatment and Labor Act), hospitals are legally required to provide emergency care regardless of your ability to pay or any billing disputes.

Important Warnings

Don't Sign Additional Waivers: Hospitals might ask you later to sign forms that waive your protections. This may be called a "Surprise Billing Protection" form. Be cautious; you don't have to sign these.

Know Your Rights Under the No Surprises Act (NSA): If insured, you're protected against surprise bills from out-of-network emergency care.

Key Points to Remember

Prepare Ahead: Carry the disclaimer written down or saved on your phone.

Stay Polite but Firm: You're protecting yourself, not refusing care.

Keep Documentation: Always keep copies or photos of your modified paperwork.

Be Persistent: If the first person can't help, ask to escalate politely.

You deserve fair treatment and fair billing. Proactively using this disclaimer can ease your stress, protect your finances, and make healthcare interactions more manageable.

REFERENCES

Becoming Climate Health Literate

1. "A Commission on Climate Change." *The Lancet* 373, no. 9676 (2009). https://www. thelancet.com/journals/lancet/article/PIIS0140-6736(09)60922-3/fulltext.

2. "Health Literacy." *Wikipedia*. Accessed March 13, 2025. https://en.wikipedia.org/ wiki/Health_literacy.

3. Vince, Gaia. *Nomad Century: How Climate Migration Will Reshape Our World*. New York: Flatiron Books, 2022.

4. United States Environmental Protection Agency. *Social Vulnerability Report*. Washington, DC: U.S. Environmental Protection Agency, 2021. Accessed March 13, 2025. https://www.epa.gov/cira/social-vulnerability-report.

Climate Change Makes Everything Worse

1. Vince, Gaia. *Nomad Century: How Climate Migration Will Reshape Our World*. New York: Flatiron Books, 2022.

Chapter 1: Heat and Health

Heat: Why Is It Getting So Hot?

1. Goodell, Jeff. *The Heat Will Kill You First: Life and Death on a Scorched Planet*. New York: Little, Brown and Company, 2023.

2. Eisenstein, Charles. *Climate: A New Story*. Berkeley: North Atlantic Books, 2018.

Heat: What Temperature Is It?

1. Lemery, Jay, and Paul Auerbach. *Enviromedics: The Impact of Climate Change on Human Health*. Lanham, MD: Rowman & Littlefield, 2017.

2. National Recreation and Park Association. "Synthetic Sports Fields and the Heat Island Effect." *Parks & Recreation Magazine*. Accessed December 29, 2024. https://www.nrpa.org/ parks-recreation-magazine/2019/may/synthetic-sports-fields-and-the-heat-island-effect/.

3. Casey, Michael. "Takeaways from AP Report on Risks of Rising Heat for High School Football Players." *Associated Press*. Accessed December 29, 2024. https://apnews.com/article/ high-school-football-heat-deaths-climate-change-072241ef3a0783f49da4bb40af39e520.

4. Bohn, Katie. "Humans Can't Endure Temperatures and Humidities as High as Previously Thought." *Penn State News*. Accessed December 29, 2024. https://www.psu.edu/news/research/ story/humans-cant-endure-temperatures-and-humidities-high-previously-thought.

Health Impacts of Extreme Heat

1. Lemery, Jay, and Paul Auerbach. *Enviromedics: The Impact of Climate Change on Human Health*. Lanham, MD: Rowman & Littlefield, 2017.

2. Vince, Gaia. *Nomad Century: How Climate Migration Will Reshape Our World*. New York: Flatiron Books, 2022.

3. Shindell, Drew, et al. "The Effects of Heat Exposure on Human Mortality Throughout the United States." *GeoHealth* 4, no. 4 (2020). Accessed December 29, 2024. https://agupubs.onlinelibrary.wiley.com/doi/10.1029/2019GH000234.

4. National Weather Service. "Heat Index." *National Weather Service*. Accessed December 29, 2024. https://www.weather.gov/arx/heat_index.

5. Children's Hospital of Orange County. "Climate Change and Children's Health: What Parents Should Know." https://health.choc.org/climate-change-and-childrens-health-what-parents-should-know/.

6. Americares. *Climate Resilience for Frontline Clinics Toolkit: Extreme Heat*. Accessed December 29, 2024. https://www.americares.org/wp-content/uploads/ExtremeHeat CompleteModule_FINAL.pdf.

7. Americares. *Toolkit: Extreme Heat*.

8. Children's Hospital of Philadelphia. "Study Shows Climate Change Will Lead to Increase in Kidney Stones." *Newswise*. January 5, 2022. https://www.newswise.com/articles/study-shows-climate-change-will-lead-to-increase-in-kidney-stones.

Chapter 2: Air Quality

Air Quality: What Are We Breathing?

1. Figueres, Christiana, and Tom Rivett-Carnac. *The Future We Choose: Surviving the Climate Crisis*. New York: Knopf, 2020.

2. California Department of Public Health. *California Wildfire Smoke Considerations*. Sacramento: California Department of Public Health, 2022. https://www.cdph.ca.gov/Programs/EPO/CDPH%20Document%20Library/CA_Wildfire_Smoke_Considerations.pdf.

3. Burke, Marshall, et al. "What We Know About the Health Effects of Wildfire." *Stanford Report*. https://news.stanford.edu/stories/2025/01/assessing-wildfire-health-risks. Accessed April 6, 2025.

4. CBC News. "Parents of B.C. Boy Who Died of Asthma Say Reliable Air Quality Monitors Could Have Saved His Life." Accessed December 29, 2024. https://www.cbc.ca/news/canada/british-columbia/parents-focus-on-air-quality-monitors-following-sons-death-due-to-asthma-1.7017126.

5. Climate Central. "Black Detroit Children with Asthma Hardest Hit by Seasonal Allergies." Accessed December 29, 2024. https://www.climatecentral.org/partnership-journalism/black-detroit-children-with-asthma-hardest-hit-by-seasonal-allergies.

6. Johnson, N. M., Hoffmann, A. R., Behlen, J. C. et al. "Air Pollution and Children's Health—A Review of Adverse Effects Associated with Prenatal Exposure from Fine to Ultrafine Particulate Matter." *Environmental Health and Preventive Medicine* 26 (2021): 72. https://environhealthprevmed.biomedcentral.com/articles/10.1186/s12199-021-00995-5.

7. Yang, T., Wang, J., Huang, J., Kelly, F. J., and Li, G. "Long-term Exposure to Multiple Ambient Air Pollutants and Association With Incident Depression and Anxiety." *JAMA Psychiatry.* https://jamanetwork.com/journals/jamapsychiatry/fullarticle/2801116.

Air Quality: What You Can Do

1. American Lung Association. "Rising to the Challenge: Caring for Patients During Climate Disasters." Accessed December 29, 2024. https://www.lung.org/blog/climate-disasters-patient-care.

2. U.S. Department of Energy. "Weatherstripping." Accessed April 8, 2025. https://www.energy.gov/energysaver/weatherstripping.

3. Consumer Reports. "Air Purifier Buying Guide." Accessed December 29, 2024. https://www.consumerreports.org/appliances/air-purifiers/buying-guide/.

4. United States Environmental Protection Agency. "DIY Air Cleaner to Reduce Wildfire Smoke Indoors." Accessed April 8, 2025. https://www.epa.gov/system/files/documents/2023-08/DIY%20Air%20Purifier%20Infographic_508%20Compliant.pdf.

5. Malin, Zoe. "The Best NIOSH-Approved N95 Masks, According to Doctors." *NBC News.* Accessed April 8, 2025. https://www.nbcnews.com/select/shopping/best-n95-masks-ncna1288842.

6. Climate Psychiatry Alliance. *Wildfire Smoke Toolkit.* Accessed December 29, 2024. https://www.climatepsychiatry.org/wildfire-smoke-toolkit.

Chapter 3: Natural Disasters

Wildfires

1. Environmental Protection Agency. "Climate Change Indicators: Wildfires." Accessed April 8, 2025. https://www.epa.gov/climate-indicators/climate-change-indicators-wildfires.

Water

1. Ready.gov. "Floods." Accessed December 29, 2024. https://www.ready.gov/floods.

2. Figueres, Christiana, and Tom Rivett-Carnac. *The Future We Choose: Surviving the Climate Crisis.* New York: Knopf, 2020.

3. Wiegandt, Klaus, ed. *3 Degrees More: The Impending Hot Season and How Nature Can Help Us Prevent It.* Springer, 2024. Accessed December 29, 2024. https://link.springer.com/content/pdf/10.1007/978-3-031-58144-1.pdf.

4. Dunn, Thom. "The Best Emergency Weather Radio." *The New York Times*. Accessed December 29, 2024. https://www.nytimes.com/wirecutter/reviews/best-emergency-weather-radio/.

5. Collins, Dave, Karen Matthews, and Michael R. Sisak. "Floods from 'Training Thunderstorms' Lead to Dramatic Rescues and 2 Deaths in Connecticut." *Associated Press*. Accessed December 29, 2024. https://apnews.com/article/connecticut-flooding-northeast-rain-c3bce519046c3ceb7f042159efe432f4.

6. Cramer, Maria. "How to Escape Your Car in a Flood." *New York Times*. February 5, 2024. https://www.nytimes.com/article/how-to-avoid-car-drowning.html.

Power Outages

1. Pacific Gas and Electric Company. "Public Safety Power Shutoffs." Accessed December 29, 2024. https://www.pge.com/en/outages-and-safety/safety/community-wildfire-safety-program/public-safety-power-shutoffs.html.

2. Dunn, Thom. "The Best Emergency Weather Radio." *The New York Times*. Accessed December 29, 2024. https://www.nytimes.com/wirecutter/reviews/best-emergency-weather-radio/.

3. Centers for Disease Control and Prevention. "Carbon Monoxide Poisoning." Accessed December 29, 2024. https://www.cdc.gov/carbon-monoxide/about/.

4. "311 (Telephone Number)." *Wikipedia*. Accessed December 29, 2024. https://en.wikipedia.org/wiki/311_(telephone_number).

Extreme Cold

1. "The Spanish Inquisition (Monty Python)." *Wikipedia*. Accessed December 29, 2024. https://en.wikipedia.org/wiki/The_Spanish_Inquisition_(Monty_Python).

2. Auger, Nathalie, Brian J. Potter, Audrey Smargiassi, Marianne Bilodeau-Bertrand, Clément Paris, and Tom Kosatsky. "Association Between Quantity and Duration of Snowfall and Risk of Myocardial Infarction." *CMAJ* 189, no. 6 (2017): E235–E242. https://www.cmaj.ca/content/189/6/E235.

3. Oklahoma State University Extension. "Winter Hydration Is Important For Good Health." Accessed December 29, 2024. https://extension.okstate.edu/articles/2024/winter_hydration.html.

4. Lane, Kathryn, Kazuhiko Ito, Sarah Johnson, Elizabeth A. Gibson, Andrew Tang, and Thomas Matte. "Burden and Risk Factors for Cold-Related Illness and Death in New York City." *International Journal of Environmental Research and Public Health* 15, no. 4 (2018): 632. https://pmc.ncbi.nlm.nih.gov/articles/PMC5923674/.

Chapter 4: Plan and Protect

Family Communication and Evacuation Planning

1. Robinson, Chelsea. "Tampa Mayor Issues Dire Warning to Residents Ignoring Evacuation Orders." *WESH,* October 8, 2024. https://www.wesh.com/article/tampa-mayor-dire-warning-milton/62539927. Accessed April 8, 2025.

Chapter 5: Training, Teamwork, and Caring

Emergency Preparedness for Pets

1. ASPCA. "ASPCA Survey Shows 83% of Pet Owners Are Impacted by Disasters, Fewer Than Half Have Preparedness Plans." Accessed December 29, 2024. https://www.aspcapro.org/news/2021/09/22/aspca-survey-shows-83-pet-owners-are-impacted-disasters-fewer-half-have.

2. American Red Cross. "Cat and Dog First Aid Online." Accessed December 29, 2024. https://www.redcross.org/take-a-class/classes/cat-and-dog-first-aid-online/a6R0V0000015EUf.html.

3. American Red Cross. "Help Keep Your Pets Safe with Free Pet First Aid App." Accessed December 29, 2024. https://www.redcross.org/about-us/news-and-events/news/2024/help-keep-your-pets-safe-with-free-pet-first-aid-app.html.

Chapter 6: Insects and Disease

Ticks

1. Bay Area Lyme Foundation. *Fast Facts: Lyme Disease.* Accessed December 29, 2024. https://www.bayarealyme.org/wp-content/uploads/2024/05/BAL_Fast-Facts_2024_ENG_REV.pdf.

Chapter 7: Food Safety

1. Centers for Disease Control and Prevention. "Burden of Foodborne Illness: Findings." *CDC,* December 15, 2010. Accessed April 8, 2025. https://www.cdc.gov/food-safety/php/data-research/foodborne-illness-burden/index.html.

2. World Health Organization. "Food Safety." *WHO,* October 4, 2024. Accessed April 8, 2025. https://www.who.int/news-room/fact-sheets/detail/food-safety.

3. Ioanes, Ellen, and Li Zhou. "Why Food Recalls Are Everywhere Right Now." *Vox.* October 23, 2024. Accessed April 8, 2025. https://www.vox.com/food/379474/mcdonalds-e-coli-boars-head-listeria-salmonella-outbreak.

4. Centers for Disease Control and Prevention. "People at Increased Risk for Food Poisoning." Accessed April 8, 2025. https://www.cdc.gov/food-safety/risk-factors/.

5. U.S. Food and Drug Administration. "Safe Food Handling." Accessed April 8, 2025. https://www.fda.gov/food/buy-store-serve-safe-food/safe-food-handling.

Chapter 8: Mental Health

Where Is My Refuge?

1. Macy, Joanna, and Chris Johnstone. *Active Hope (Revised): How to Face the Mess We're in with Unexpected Resilience and Creative Power*. Novato, CA: New World Library, 2022.

2. Amsen, Eva. "The Art and Science of Coping With Climate Change Around the World." *Forbes*. July 7, 2023. https://www.forbes.com/sites/evaamsen/2023/07/07/the-art-and-science-of-coping-with-climate-change-around-the-world/.

3. Nelson, Kristi. *Wake Up Grateful: The Transformative Practice of Taking Nothing for Granted*. New York: Storey Publishing, 2020.

4. Amer, Aïda, and Tiffany Herring. "Channeling Climate Anxiety into Action." *Axios*. November 6, 2024. Accessed April 8, 2025. https://www.axios.com/2024/11/06/climate-change-anxiety-youth-action.

Chapter 9: Signals and Sirens

Uses of Social Media During Extreme Weather

1. Lam, N. S. N., Meyer, M., et al. "Improving Social Media Use for Disaster Resilience: Challenges and Strategies." *International Journal of Digital Earth* 16, no. 1 (2023): 3023–3044. https://doi.org/10.1080/17538947.2023.2239768.

Information and Disinformation

1. Figueres, Christiana, and Tom Rivett-Carnac. *The Future We Choose: Surviving the Climate Crisis*. New York: Knopf, 2020.

2. Ballard Brief. "The Effects of Medical Misinformation on the American Public." Accessed December 29, 2024. https://ballardbrief.byu.edu/issue-briefs/the-effects-of-medical-misinformation-on-the-american-public.

3. Vosoughi, Soroush, Deb Roy, and Sinan Aral. "The Spread of True and False News Online." *Science* 359, no. 6380 (March 9, 2018): 1146–1151. https://pubmed.ncbi.nlm.nih.gov/29590045/.

4. Ballard Brief. "The Effects of Medical Misinformation on the American Public." Accessed December 29, 2024. https://ballardbrief.byu.edu/issue-briefs/the-effects-of-medical-misinformation-on-the-american-public.

5. Borges do Nascimento, Israel Júnior, et al. "Infodemics and Health Misinformation: A Systematic Review of Reviews." *Bulletin of the World Health Organization* 100 (2022): 544–56. https://pmc.ncbi.nlm.nih.gov/articles/PMC9421549/.

Chapter 10: We Need to Talk:
Health Insurance 101 for Climate Emergencies

After the Disaster: Health Insurance Claims

1. Consumer Reports. "How to Lower Your Medical Bills." Accessed April 8, 2025. https://www.consumerreports.org/money/medical-billing/how-to-lower-your-medical-bills-a3379293456/.

2. Allen, Marshall. *Never Pay the First Bill: And Other Ways to Fight the Health Care System and Win*. New York: Portfolio, 2021.

3. U.S. Department of Health and Human Services. "File a Complaint if Denied Access to Medical Records." Accessed April 8, 2025. https://ocrportal.hhs.gov/ocr/smartscreen/main.jsf.

RECOMMENDED READING

Disasterology: Dispatches from the Frontlines of the Climate Crisis

by Samantha Montano

A blend of memoir and expert analysis examining our world's growing vulnerability to natural disasters and how we can better prepare.

How to Prepare for Climate Change: A Practical Guide to Surviving the Chaos

by David Pogue

A comprehensive, actionable guide to preparing yourself and your community for the challenges of our changing climate.

The Atlas of a Changing Climate: Our Evolving Planet Visualized with More Than 100 Maps, Charts, and Infographics

by Brian Buma

A visually stunning resource that helps readers understand complex climate data through accessible maps and graphics.

Enviromedics: The Impact of Climate Change on Human Health

by Jay Lemery and Paul Auerbach

An essential examination of the direct connections between climate change and human health outcomes.

Climate Optimism: Celebrating Systemic Change Around the World

by Zahra Biabani

An inspiring look at positive climate action and solutions being implemented globally.

ACKNOWLEDGEMENTS

To Darragh O'Carroll MD, thank you for your early encouragement.

To my *On Deck Climate Tech* cohort: Wassa Cisse, Patrick Beckley, Saleha Shah, Tanya Meshram, and Mustafa Ozer.

To Candice Ammori, I'm grateful for the introduction to a world of purpose and possibility. You showed me that climate innovation is not just necessary, but achievable.

To Jim Gentz, Shaquana Smiley, Gavin Maloney, and Deidre Crawford. Your encouragement, creativity, and willingness to engage in big ideas helped shape my thinking and this work.

Thank you to The Weather Channel for your collaboration.

To my meditation group: Luciano Pinheiro, Hanna Sone, and Karen Chen. To Emma Swan and John Brown for your vision.

Special thanks to the brilliant editorial and design team who brought this book to life. Brandi Lai, for your insightful developmental editing. Lily Capstick, for your precise and thoughtful copyediting. Nelly Murariu, for your elegant typesetting and beautifully crafted cover design.

To the PYP team: Jenn Grace, Catherine Whiting, and Alex Loutsenko, thank you for your energy, dedication, and the clarity you bring to meaningful work. Your support made all the difference.

Thank you to Grace Farms for a serendipitous meeting that shifted the course of my life and opened new paths. Thank you, Frank Kwei, for introducing me to the world of tea.

To Felicia Rubinstein, Lisa Miniter, and the HAYVN community, thank you for your early support and belief in this work, long before it had a title or shape. Your encouragement mattered.

To my friends John Mohring, John Curtin, Carolyn McKenna, Tabitha Shustock, Tom Butler, Janice and Pete Olson, your presence and steady support remind me why this work matters and who I do it for.

To Rich Cimaglia for the porch and the sandhill cranes. Your memory lives on in us.

To the libraries of Westport, Darien, and New Canaan, Connecticut. Thank you for the quiet corners. So much of *this* book was born between your shelves.

To the Futurists who helped me stretch my vision and sharpen my thinking: Frank Spencer, Sohail Inayatullah, Jane McGonigal, and Bronwyn Williams.

To Matt Selznick, thank you for bringing samcherubin.com to life.

STAY CONNECTED

Thank you for reading *The Crisis Is Here*. This is just the beginning of our journey toward health resilience in a changing climate.

Visit my website to find additional resources, updates and Climate Cat at samcherubin.com.

Join my community: Subscribe to my newsletter for exclusive content and timely climate health alerts at samcherubin.com/subscribe.

Follow Climate Cat and me on social media:

in https://www.linkedin.com/in/samcherubin/

🦋 @samcherubin.bsky.social

Bring this message to your organization: Book me for speaking engagements, workshops, or consulting on climate health preparedness at samcherubin.com/contact.

ABOUT SAM CHERUBIN

Sam Cherubin is a Connecticut-based author, product manager, and futurist whose work focuses on climate change, health, and innovation. With over twenty years of experience managing award-winning digital platforms in the health insurance industry, Sam brings a blend of systems thinking, storytelling, and strategy to the most urgent challenges of our time.

Drawing on his experience in health innovation and climate resilience, Sam offers practical guidance for navigating extreme weather and everything it brings. The world we took for granted—reliable weather reports, government support, accessible science—is being taken away. Sam's work is about preserving our lives and communities when the ground shifts beneath us.

When he's not designing digital solutions or collaborating with Climate Cat, Sam enjoys meditating, writing poetry, and being outdoors. His work is grounded in compassion and pragmatism: a belief that while the future will be more chaotic than we imagine, it must be met with clarity, courage, and care.

Learn more about Sam's work and connect at samcherubin.com.

ABOUT CLIMATE CAT

Climate Cat was born when the skies turned orange, and the power grid started to fail. He heard Nature's cry for justice and the mourning song of what we've lost.

He is a responder, a witness, a bridge. He remembers the world before it cracked: rivers that ran clear, silence without smoke, children who could play outdoors.

Memory is his compass. He watched the wild things disappear, people get sick, and storms form and attack.

Climate Cat fights the systems that criminalize healing. That outlaw the truth. He navigates emergency rooms without power, shelters without air filters, communities without access to care.

He has become a scholar of the Anthropocene, fluent in heat domes, vector-borne illness, and the language of smog.

But he's also a poet, carrying stories from elders and whispers from the trees. He knows that resilience is relational, and survival is a collective act.

In every smoky horizon, he looks for light. In every displaced person, he sees the whole planet. Climate Cat is here to prepare us for the work.

Because Liberty carried the torch.

And Climate Cat is Liberty's knight.

Passing the Torch

AFTERWORD

By Samantha Taylor,
Founder, Reputation Dynamics and Elephant Art Shop

We often overlook a key ally in protecting public health and stabilizing our climate: trees.

Trees are among the most immediate, accessible, and cost-effective climate and health solutions we have. Often called the lungs of our Earth, trees absorb carbon dioxide, release oxygen, and help regulate local temperatures, directly benefiting the communities who live among them.

Urban tree cover cools neighborhoods, reducing the urban heat island effect and lowering the risk of heat-related illnesses. Forests filter air pollutants that contribute to asthma and cardiovascular disease, especially in children and the elderly.

Despite their value, we continue to degrade and destroy forests at alarming rates. In 2023 alone, we lost over 10 million hectares of tree cover in tropical regions—nature's lungs and our frontline defense. When forests are lost, it's not just carbon that's released into the atmosphere. It's resilience: our ability to adapt and thrive that disappears.

In a warming world, trees aren't just symbolic. They're strategic. They are under threat and key to climate mitigation.

For families and communities looking for near-term action, trees offer a tangible, scalable fix. Whether through planting efforts, preserving green space, or supporting reforestation, we have the power to restore this critical layer of protection—one that mitigates climate change while strengthening public health.

Hope

THE CRISIS IS HERE

THE B CORP MOVEMENT

Certified

Dear Reader,

Thank you for reading this book and joining the Publish Your Purpose community! You are joining a special group of people who aim to make the world a better place.

Corporation

What's Publish Your Purpose About?

Our mission is to elevate the voices often excluded from traditional publishing. We intentionally seek out authors and storytellers with diverse backgrounds, life experiences, and unique perspectives to publish books that will make an impact in the world.

Beyond our books, we are focused on tangible, action-based change. As a woman- and LGBTQ+-owned company, we are committed to reducing inequality, lowering levels of poverty, creating a healthier environment, building stronger communities, and creating high-quality jobs with dignity and purpose.

As a Certified B Corporation, we use business as a force for good. We join a community of mission-driven companies building a more equitable, inclusive, and sustainable global economy. B Corporations must meet high standards of transparency, social and environmental performance, and accountability as determined by the nonprofit B Lab. The certification process is rigorous and ongoing (with a recertification requirement every three years).

How Do We Do This?

We intentionally partner with socially and economically disadvantaged businesses that meet our sustainability goals. We embrace and encourage our authors and employee's differences in race, age, color, disability, ethnicity, family or marital status, gender identity or expression, language, national origin, physical and mental ability, political affiliation, religion, sexual orientation, socio-economic status, veteran status, and other characteristics that make them unique.

Community is at the heart of everything we do—from our writing and publishing programs to contributing to social enterprise nonprofits like reSET (www.resetco.org) and our work in founding B Local Connecticut.

We are endlessly grateful to our authors, readers, and local community for being the driving force behind the equitable and sustainable world we are building together.

To connect with us online or publish with us, visit us at www.publishyourpurpose.com.

Elevating Your Voice,

Jenn T. Grace
Founder, Publish Your Purpose

CLIMATE CAT™

www.ingramcontent.com/pod-product-compliance
Lightning Source LLC
Chambersburg PA
CBHW060357200326
41518CB00009B/1176